TECHNIQUES
OF
BIOLOGICAL PREPARATION

John Simpkins
B.Sc., Dip.Ed., M.I.Biol.
Lecturer in Biology, People's College of Further Education, Nottingham

Glasgow BLACKIE London

BLACKIE & SON LIMITED
Bishopbriggs
Glasgow G64 2NZ

5 Fitzhardinge Street
London W1H 0DL

First published 1974

ISBN 0 216 89767 X

Printed in Great Britain by
Thomson Litho, East Kilbride, Scotland.

PREFACE

The teaching and study of Biology involves constant reference to preserved specimens of organisms once living. Either the whole organism or part of it may be treated and arranged to display particular features to best advantage. Effective study of anatomical characteristics requires well-prepared and accurate specimens, and it is the object of this book to present some of the techniques which may be employed in the preparation of such specimens.

The methods described by no means constitute an exhaustive list, but are meant to form a basis for anyone intending to engage in such work. The book should prove particularly useful to Biology laboratory technicians and teachers of Biology at all levels. It is of advantage to prepare one's own specimens for one's own special purposes; set against the cost of specimens prepared commercially, the saving can be enormous. Many Biology students would understand the topics studied by preparing their own specimens, and the techniques are presented in such a way as to be easily followed and understood by all. It is hoped that some may find these techniques the beginning of a pleasurable and satisfying pastime.

J.S. NOTTINGHAM 1974

CONTENTS

COLOURED PLATES

SECTION A

PREPARATION OF WHOLE SPECIMENS

CHAPTER ONE

PRESERVATION

A. WET PREPARATIONS

WITH FEW EXCEPTIONS, MOST BIOLOGICAL SPECIMENS THAT ARE to be preserved and prepared for display require initial treatment with a fluid *fixative*. Immediately an animal dies or a plant is removed from its substrate, the tissues will start to degenerate and the form will distort. There are several reasons for this. Moisture will evaporate readily from specimens exposed to the air. Plants are particularly susceptible to this, as tissue turgor is more important to them as a skeleton than in animals. Placing a fresh specimen into a fluid preservative overcomes this problem.

All dead tissues are susceptible to infestation by a number of micro-organisms, particularly bacteria and fungi. Animal carcasses are specially liable to rapid decay from within, where bacteria normally inhabiting the gut invade the other organs. For this reason, all larger animals should be incised in their abdominal body wall to allow quick infusion of the fixative.

All cells seem to digest themselves following death of the body to which they belong. It seems that some mechanism triggers the release of hydrolytic enzymes within the cells, which then dissolve. Post-mortem distortion and decay begin surprisingly rapidly and, if specimens are to be preserved, immediate fixation following the death of the specimen is essential.

Many fluids have fixative properties, but the two most

widely used are *ethyl alcohol* and *formalin*. Each of these has particular properties, and the choice of fixative depends very largely upon the specimen being treated. Alcoholic fixatives which penetrate the tissues rapidly yet tend to cause shrinkage are recommended for herbaceous plants and most small invertebrates.

These fixatives would normally contain 70–90% alcohol. Formalin penetrates tissues less quickly, but causes less distortion and should be used with fleshy succulent plants and vertebrates. Formalin is obtained commercially as a 40% aqueous solution of formaldehyde. This should be diluted for fixation to give a solution of up to 5% formalin (i.e. 2% formaldehyde). The formulation for a general plant fixative to cover most circumstances is as follows:

Alcohol	15 volumes
Formalin	1 volume
Glacial acetic acid	1 volume
Distilled water	10 volumes

One major disadvantage of wet preparations is that colours are rarely preserved, especially in plants. Reference is made in Chapter Two to several solutions that may be used to preserve colour in plants prior to embedding in clear polyester resin. The same problem applies to animal specimens and does not seem to have been satisfactorily solved.

Mounting specimens for display

Specimens must be attached to a backing plate cut to fit exactly into an appropriate museum jar. Several methods are available. The main difficulty is to arrange the specimen securely, so that it will remain in place when the jar is handled. On the other hand, wet preparations tend to become very brittle and are easily damaged if secured too tightly. Specimens are usually tied onto the back plate with thread or, better, nylon monofilament. Suitable holes must be drilled beforehand, and the specimen is laid on the back plate to locate the best site for these. Some plant material that is not too bulky may be stuck onto the plate with an adhesive that will not dissolve

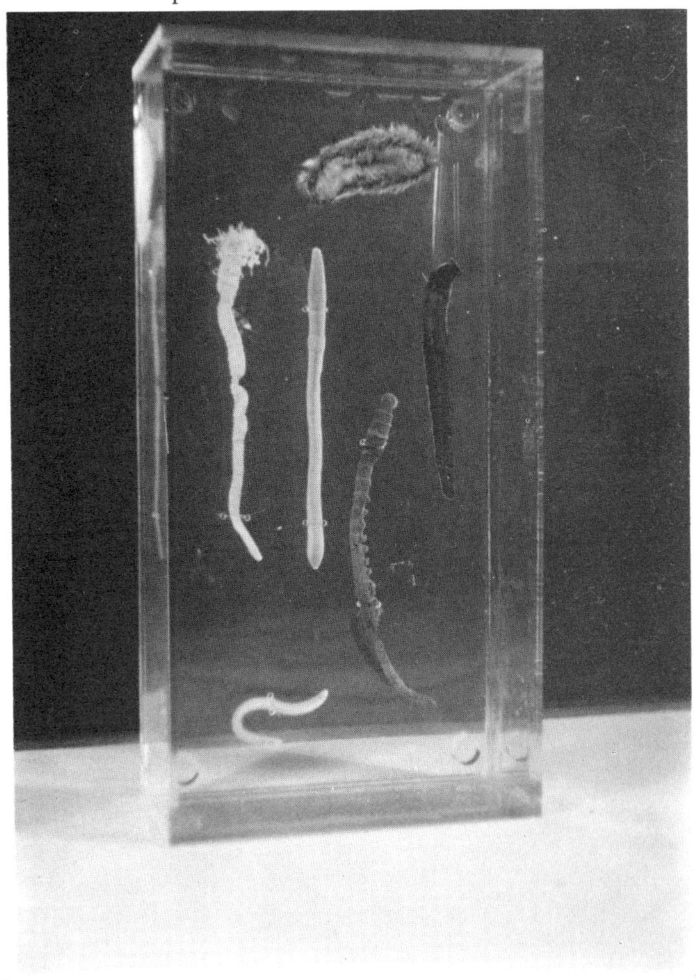

Fig. 1:1 Formalin-preserved *Annelids*

in the preservative used. Neater presentations are obtained this way, but where filaments must be used, the knots tied should be behind the back plate and not visible.

Wherever possible, Perspex museum jars are recommended. They are lighter, more easily handled and generally of better appearance than glass. Another advantage of Perspex jars is that their lids may be more neatly and effectively sealed with Perspex cement. However, if alcoholic preservatives are used, glass jars are necessary as Perspex is attacked by alcohol. Figure 1:1 shows several formalin-preserved specimens displayed in a Perspex container.

B. DRY PREPARATIONS

Some animals and many plants have bodies which may be preserved, mounted and displayed in a dry state. This has certain advantages in that the surface details of a specimen are likely to be well preserved, and the techniques are fairly straightforward and easy to perform. On the other hand finished specimens are often fragile and, when handled and stored, they are liable to be damaged. Nevertheless, in certain circumstances the following methods may be desirable.

Pressing

Most plants may be effectively displayed in a dry pressed state. The specimen is arranged on a sheet of drying paper in such a way that the features to be displayed will not be hidden or damaged by pressing. A little experience is necessary to achieve this. Many specimens may be stacked in a pile and a weight applied to the top to press them. Each specimen must be separated from the drying paper above it by a sheet of grease-proof paper. This prevents the specimens from being pulled off

Plate 1 Pressed seaweeds

their own backing paper by the drying sheet above when they are removed after pressing.

In order to effect quick drying and to prevent fungal infestation, the drying papers should be changed daily and the whole process carried out in a warm place. Special pressing apparatus may be made or purchased, usually consisting of two outer frames which are strapped together, pressing the drying papers in between. (See figure 1:2.)

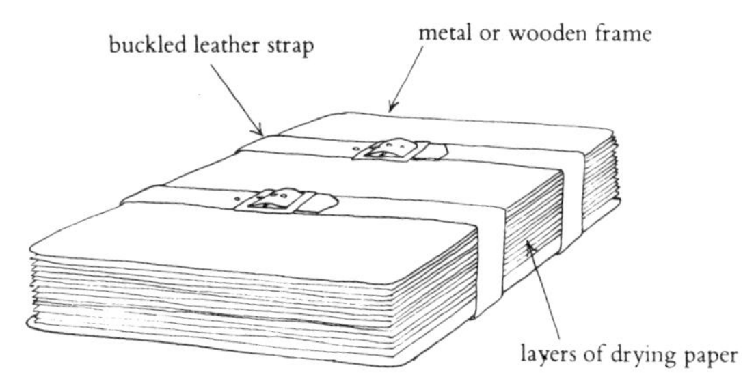

Fig. 1:2 A simple plant press

Some particularly bulky plants present problems when pressing, where the tissues may be squashed and rendered useless. This may be overcome by providing a base into which the plant may partially sink while it dries and the less bulky parts press. (See figure 1:3.)

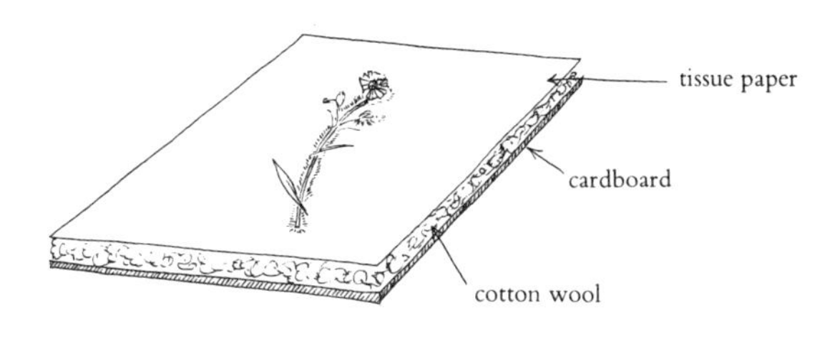

Fig. 1:3 Method for pressing bulky specimens

When pressing and drying are complete, the specimen is attached to a sheet of backing paper on which it will be displayed and stored in an herbarium. Biological supply companies provide a range of herbarium sheets for this purpose. The choice of materials for sticking the specimen onto the paper depends upon personal preference and the size of the specimen. Smaller plants may be stuck by applying suitable quantities of special glue to the parts of the plant which contact the paper.

Botanical Mounting Paste may be obtained from Watkins and Doncaster Ltd., Welling, Kent. While the final appearance may not be as good, it may be necessary to stick larger specimens with strips of adhesive tape over the stem or other parts or even with fine thread for very bulky plants. A label should always be added to the paper with such information as is shown in figure 1:4.

Herbarium number ______ date ______
Genus ______ species ______
Common name ______
Classification ______
______
______
Locality ______ Collector ______
General remarks ______
______

Fig. 1:4 A typical herbarium label

Pressed specimens must be stored in drawers with ventilation adequate to prevent the specimens becoming damp or infested with pests. Most biological suppliers offer special herbarium cabinets for those intending to make a collection.

Aquatic plants may also be pressed and dried. They must first be arranged on backing paper and this may be done by the method illustrated by figure 1:5. Once the specimen is removed from the water, it is dried and pressed in the usual way. The author has tried this technique successfully with a variety of small seaweeds, pressing them between sheets of blotting paper. The results, shown in plate 1 (page 4) have faded somewhat over the years and specimens are best stored in the dark. Plate 2 (page 21) shows a pressed terrestrial plant.

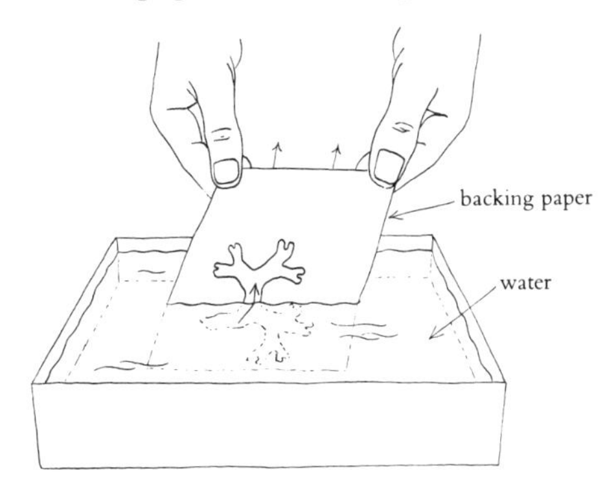

Fig. 1:5 Arranging a marine alga on backing paper

Pinning

Insects are usually best displayed by pinning them out in such a position that most of their anatomical features can be seen easily. The specimen must be handled carefully and, if freshly killed, pinned out as soon as possible. If the animal has stiffened in a contorted position, it must be relaxed before any attempt is made to reposition the wings and limbs. Otherwise these will easily snap off.

Biological suppliers advertise special insect-relaxing fluids, but in practice a 50% aqueous solution of glycerol is ideal. In a closed vessel, this provides an atmosphere of a humidity just right for relaxing specimens. Figure 1:6 illustrates a typical arrangement that might be used.

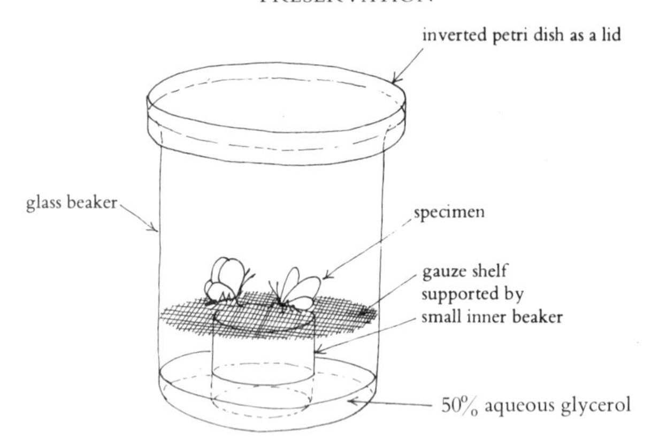

Fig. 1:6 Relaxing insects prior to pinning out

Pinning out and setting specimens in the required position is a skilled operation requiring much patience and experience. A variety of different types of *setting board* is available. One should be selected that allows the thorax and abdomen of the specimen to be inserted into the central groove without difficulty while the wings lie flat on the plateaux. Figure 1:7 shows a specimen in place on a setting board. The pin used to secure the specimen is driven through the thorax of most insects. The wings are then held in place by flat strips of card

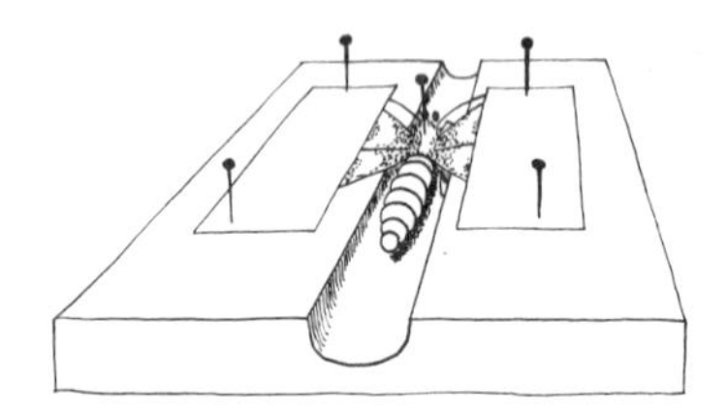

Fig. 1:7 A specimen on the pinning board

or paper pinned over the wings as shown in figure 1:7. In order to move the wings, the head of a pin will prove useful, hooked over the large *vein* on the leading edge of the wing. Do not use a sharper instrument or the wing will rip. Do not use your fingers or you may damage the patterns of pigment on the wings' surfaces.

The positioning of the wings is important, as the object is to expose as much of the specimen's features, as well as to obtain an attractive display. Conventionally, the trailing edge of the forewing should be at right angles to the axis of the body and should just overlap the leading edge of the hindwing. (See plate 3, p. 36.) Some specimens have such broad or narrow wings that this arrangement would have to be modified.

When the specimen has set (it should be left for several days) it can be pinned into a display cabinet and supplied with a label. It is conventional to push the specimen's pin through the label which will be held a little distance below the specimen.

Some specimens, particularly the bulkier forms, tend to discolour with time due to the fat and grease in their bodies. This may be cured by soaking such a specimen in a series of dishes of *benzole* until that fluid is no longer discoloured by the specimen.

Particularly small specimens which would be destroyed by pinning may be glued directly onto a backing card which may

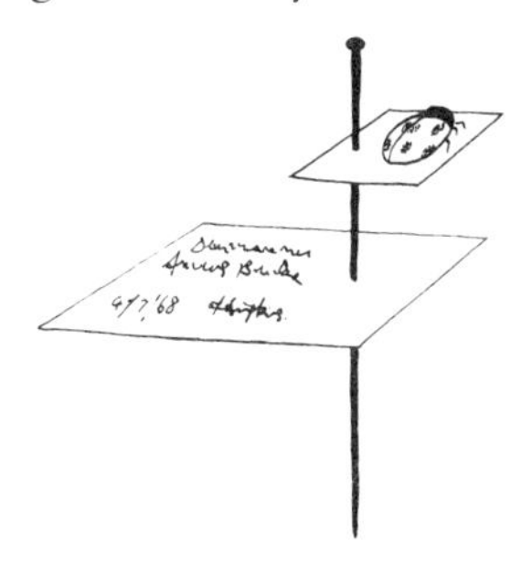

Fig. 1:8 Method for mounting a small insect

be made from clear celluloid if the underside of the specimen is important. This support may now be pinned to a suitable label and the whole thing put into a display cabinet. (See figure 1:8.)

Some taxonomic and genetic studies have involved the *venation* of the wings. To show this, a specimen is prepared in the usual way but with one of its forewings removed.

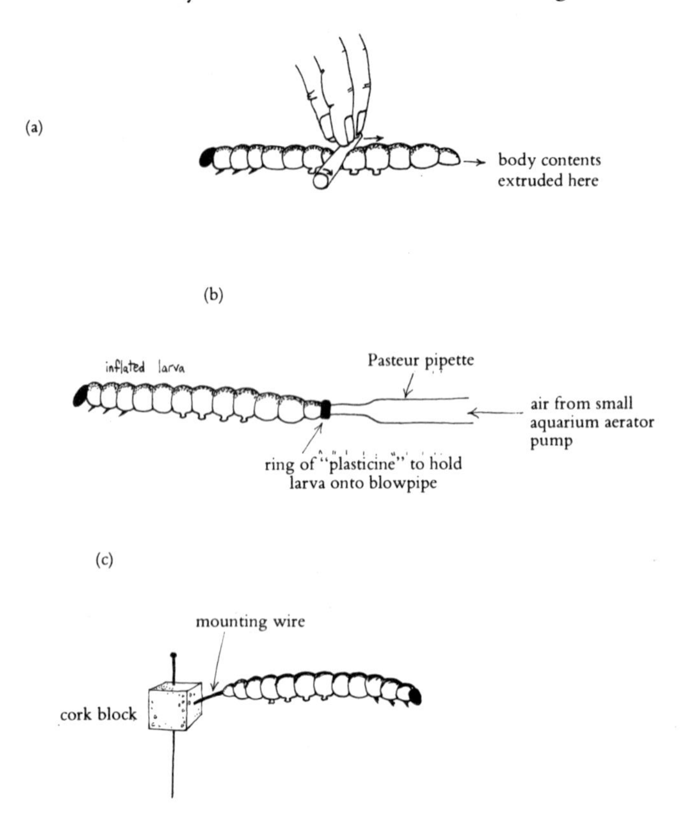

Fig. 1:9 Preparing and mounting a caterpillar

This is treated to remove its scales and pigmentation by gently brushing its surface with a fine brush while held beneath a bleaching agent such as dilute *Domestos.* After a thorough wash in water, and a dip in alcohol to help dehydration, the wing is blotted and then dried in air. Now the cleaned wing may be stuck onto a backing card and the specimen to which it belongs may be pinned through that.

Larvae may be dried and mounted when their body contents have been removed. This is a very delicate operation and requires much practice. Lay the specimen onto a clean flat tile and place a round-shafted pencil or glass rod about one third the way up its body from the hind end. Press firmly, and slowly roll the rod down towards the anus, through which it should be possible to remove the contents from within. Then start a little nearer to the head and repeat the procedure. When only the skin remains, this must now be dried in its original shape and size. A small pipette may be introduced into the anus and the body held onto this. Now blow gently through the pipette while holding the whole preparation over a hot plate. In a few minutes, when the inflated specimen dries, it may be mounted. This may be either on a wire pushed through the anus into the body cavity or directly stuck onto a support card. Figure 1:9 summarizes the whole procedure.

As with the storage of pressed plants, infestation may be a problem and suitable boxes must be used. These are most often cork- or foam-backed thick cardboard boxes with glass lids. They may be obtained in a variety of sizes from any biological supplier.

Wax impregnation

This technique developed by Hochstetter involves treating a fixed animal in such a way that all its body fluids are replaced and tissues impregnated with paraffin wax.

A vessel of a suitable size to contain the specimen in a life-like position must first be obtained. Once the specimen is fixed, it will be stiff and must therefore be placed in fixative in the position required at the end of the process. Fixation is effected by immersion for several days in 5% formalin. Even with small specimens, say of goldfish size, it is advisable also to inject the body with this fluid.

The carcass must now be dehydrated by immersion in increasing strengths of alcohol, say 50%, 70%, 90% and then two treatments in absolute alcohol. Again inject the body as well as immersing it, and leave in each treatment for at least 24 hours. Now *clear* the specimen. (This is a term derived from histological technique and does not necessarily mean that the tissues will become transparent.) Clearing is effected by injection with and immersion in chloroform for three or four days.

The next step is gradually to impregnate the tissues with paraffin wax, which will replace the chloroform used to clear. Prepare a mixture of molten paraffin wax, saturated with chloroform. It must be realized that chloroform will quickly evaporate at the temperature required to melt the wax (55°–60° C) and this stage must be carried out in a fume cupboard. It must also be emphasized that any flame used to heat up the wax be kept well away from chloroform fumes for obvious reasons. If the molten wax is added to the chloroform rather than vice versa, a layer of unmixed chloroform may remain in the bottom of the vessel. Provided a large proportion of the specimen is not projecting into this, it does not seem to interfere with the impregnation. Adding chloroform to molten wax is liable to drive off most of the former before it has a chance to mix, unless it is pipetted into the depths of the wax in the vessel (use a safety pipette!) A close-fitting lid must now be placed on the vessel to retain the chloroform while the specimen is incubated at 37° C. Removal of the lid after a few days will allow most of the chloroform to be driven off at that temperature.

Now transfer the partially impregnated carcass to a vessel containing fresh paraffin wax kept at a temperature a couple of degrees above its melting-point. After a week or so, the specimen may be removed and any excess wax allowed to drip off. Careful wiping of the surface with a wad of cotton wool slightly moistened with xylene will remove any surface wax and give the skin a clean appearance. Clean any hair on the specimen with a brush dipped in xylene. Xylene is an obnoxious liquid; care should be taken not to allow it to contact one's own skin or its vapour to reach the eyes or nose. Figure 1:10 shows a European pond tortoise, *Emys orbicularis,* treated in this way. Such specimens are very easily handled and stored, and after some practice may be easily prepared.

Fig. 1:10 Wax-impregnated *European pond tortoise*

References

Mahoney, R. (1966) *Laboratory Techniques in Zoology,* Butterworths.

Purvis, M. J., Collier, D. C., and Walls, D. (1966) *Laboratory Techniques in Botany,* 2nd. Ed., Butterworths.

Wagstaffe, R. and Fidler, J. H., *The Preservation of Natural History Specimens,* Vols. I and II (1955 and 1968) Witherby.

CHAPTER TWO

RESIN EMBEDDING

THE TECHNIQUE OF DISPLAYING PRESERVED BIOLOGICAL SPECIMENS embedded in clear polyester resin blocks has many advantages compared with the more traditional methods. Otherwise very delicate specimens can be handled, studied and stored easily with little fear of damage. Modern resins maintain their clarity well and their optical properties will even allow low-power microscopic study of the specimens embedded in them. Resin embedding dispenses with the need to use unpleasant preservatives, such as formalin, which evaporate and often eventually leak from even the best of containers. Indeed many materials, particularly botanical specimens, are difficult to preserve adequately for display purposes (especially their colour) other than by freeze-drying or photography. The first of these alternatives is expensive and rarely feasible, the second is a poor substitute for the real thing.

With careful selection and handling of materials and a little practice, many excellent preparations can be made.

Preparation of specimens to be embedded

Some specimens, especially insects, require no special preparation and can be embedded dry. They must be arranged in a suitable position (chapter one, page 7, for dry insect-mounting techniques) and all traces of moisture eliminated from them. Pigmented butterfly and moth wings, however, will discolour and become transparent when they contact the resin. One suggestion to overcome this is to apply several layers of hair lacquer gently from an aerosol spray prior to embedding in resin. This has proved effective for a variety of specimens. Some specimens may have grease on their surfaces, such as the cuticle of large insects. This must be removed before embedding by careful application of alcohol to the specimen's surface with cotton wool.

Most specimens, however, will require some fixation in liquid before embedding. A good general technique is to immerse the material in a solution of isopropyl alcohol (propan–2–ol), say 30% upwards, for a few days. Then remove excess fluid from the surfaces by gently blotting and embed immediately. Specimens left too long after removal from this fixative will allow resin to enter through their skins and become transparent. Skeletal transparencies such as those stained with Alizarin Red (see chapter eight, page 72) may be prepared for embedding by treating them with increasing strengths of ethylene glycol solution, say 50%, 70%, 90%, then two changes of 100%. The time for immersion of a specimen in each solution depends very largely on its size, but after some experiment, good results can be obtained.

Plant specimens are more difficult to prepare for resin embedding, especially in order to preserve their colour. Several solutions have been concocted for coloured flowers. The basis of these is:

Tertiary butyl alcohol	100 g
Thiourea	1 g

For red or pink flowers add 2 g citric acid; for blue or green flowers add 2 g sodium citrate. These solutions may be appropriately mixed for flowers of intermediate colour. Specimens should be soaked in one of these for about a day. Keep them at say 30° C, as the butyl alcohol solidifies at about 26° C. They are then dried in an incubator set at 40–50° C. Great care should be taken when handling, as plants become very delicate after this treatment, and embedding should follow immediately.

Yellow flowers also benefit from this treatment. Because of the fragility imparted to the specimens by this treatment they should be arranged in the appropriate fluid in the position required in the final cast.

The green colour of plant tissues not treated as above may be preserved as follows:

Prepare a 20% solution of glacial acetic acid that has been saturated with copper acetate. Place a beaker of this in a water bath kept at 100° C and immerse the fresh material in the hot acid solution. When the green colour returns after initially browning, the material is removed and washed in 50% isopropyl alcohol. Embedding should follow immediately after drying by gently blotting.

Making a resin block

Most resins available commercially for this purpose require the addition of a catalyst and an activator before they will set to a hard clear block. The setting or *curing* process is exothermic, the heat liberated distorting and cracking a block that is cast too thick. Consequently the technique of casting a block in layers must be employed, pouring several layers of resin each thin enough not to distort on curing, and each poured when the previous layer has set. Some manufacturers supply a preactivated resin to which only a catalyst need be added to effect the cure. This may be used to prepare larger blocks in one piece before the need for layering. To a certain extent exothermic distortion may be minimized if the cure is slowed down by adding less catalyst or if the block is kept cool.

However, a block cast in two or three layers is not unsightly and, if finished off properly, the layering is hardly visible at all. Indeed it is specifically desirable in the case of some dry specimens to half cover the material with one resin layer first before completing the block with a second layer. In this way most of the air trapped in the specimen will be driven off into the atmosphere before the second layer is poured. If covered in one pouring, the exothermic cure will drive the air out of the specimen and into the resin. After setting, the layer of air around the specimen will remain as a permanent 'silvering', masking all surface features.

The amounts of resin, activator and catalyst used and the proportions in which they are mixed depends largely upon the brand of goods obtained. Suitable directions should be provided by the manufacturers on request. *Trylon* resins have proved most satisfactory in the following proportions:

Resin EM 301	6 oz
Catalyst paste	8 g
Liquid activator	40 drops

or

pre- activated Resin EM 301 PA	6 oz
Liquid catalyst	108 drops

Resins are best mixed in disposable greased paper cups that can be discarded afterwards. Great care should be exercised when handling these materials, as they are highly inflammable, and if added together in the wrong order can produce an explosive mixture. The manufacturer's directions should be followed at all times.

The material from which the mould is made is very largely a matter of choice and availability. A glass mould will give a smooth finish to the resin cast in it, but will require pre-treatment with a releasing agent (any non-siliconized wax polish) to allow removal of the block after curing. The best material for moulds is polythene. This is not attacked by resin, gives a good surface to the block, does not require a releasing agent, and is sufficiently flexible to be manipulated when releasing the block. Cleaned washing-up-liquid containers are excellent. Dust must not be allowed to settle on castings while the resin is curing, and they must be put in a safe place and covered during this stage. Using the proportions given above, setting should be complete in 4 to 6 hours. The sequence of illustrations given in figure 2:1 summarizes the stages of preparing a resin block containing a suitably prepared specimen.

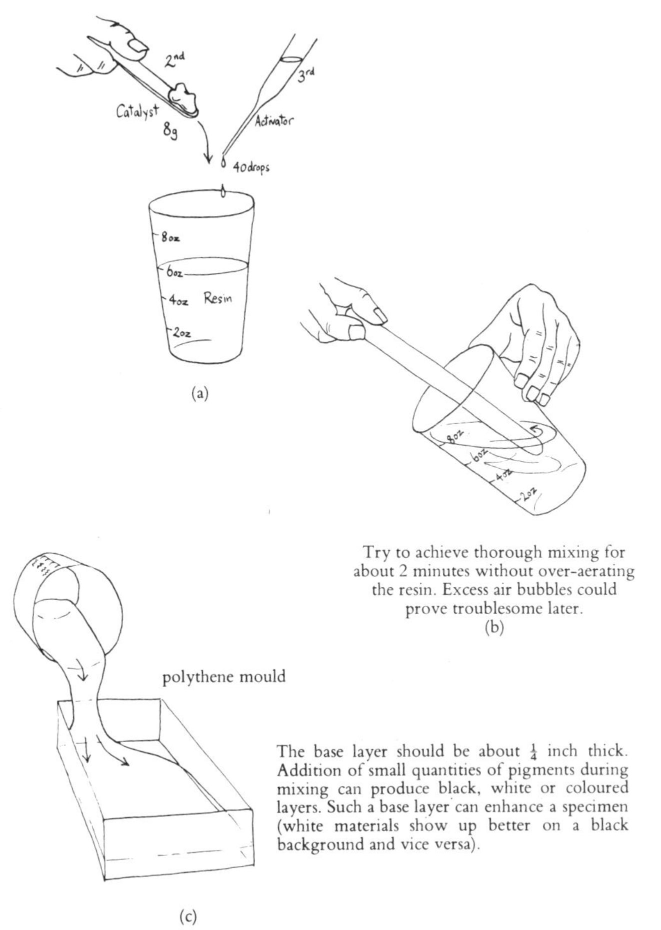

Fig. 2:1 Steps in the preparation of a seahorse embedded in clear polyester resin

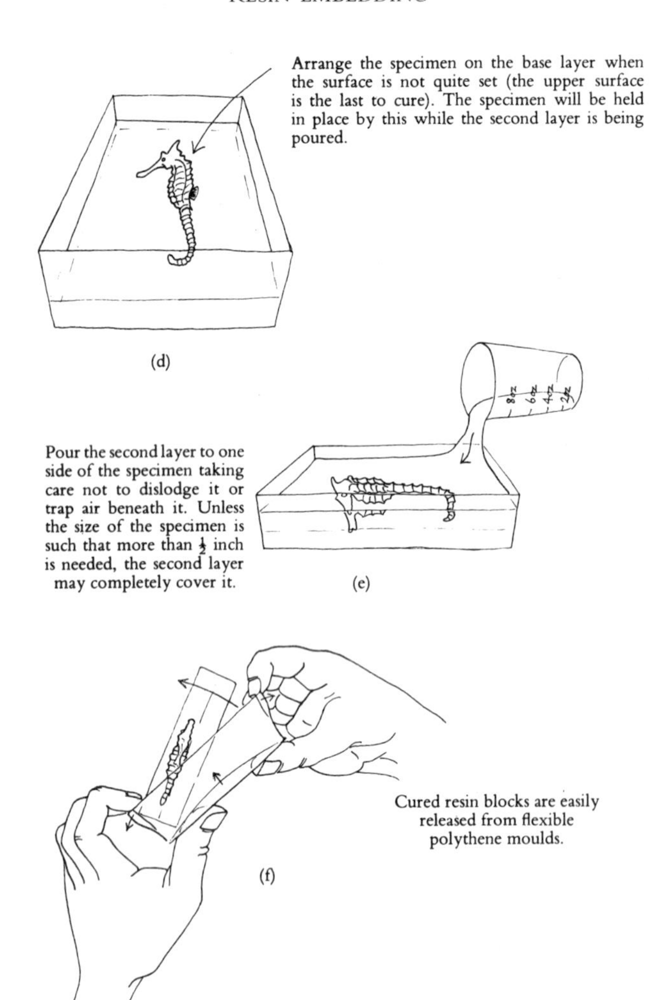

Arrange the specimen on the base layer when the surface is not quite set (the upper surface is the last to cure). The specimen will be held in place by this while the second layer is being poured.

(d)

Pour the second layer to one side of the specimen taking care not to dislodge it or trap air beneath it. Unless the size of the specimen is such that more than ½ inch is needed, the second layer may completely cover it.

(e)

Cured resin blocks are easily released from flexible polythene moulds.

(f)

Plate 2 Pressed terrestrial plant

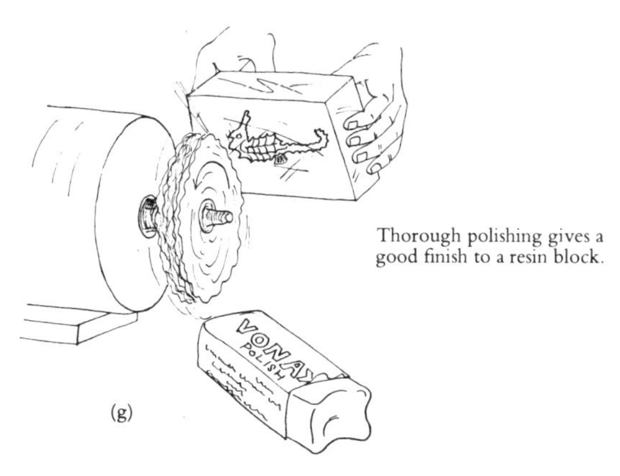

(g) Thorough polishing gives a good finish to a resin block.

Polishing and finishing

When curing is complete, the surface of a resin block will still be somewhat tacky. This must be removed before polishing. A quick wash in acetone will do this, although sometimes it has proved necessary only to press a thin polythene sheet onto the surface and peel it off again. The surface of the block can now be polished. Fine-grade wet-and-dry emery paper may be used initially, finishing off with a metal polish such as 'Brasso' applied on a soft dry rag. If a lathe is available, polishes such as 'Vonax' can be applied on a soft mop with considerable saving of muscular effort, as well as providing a better finish.

Pieces of felt may be stuck to the surfaces of the block so that the resin is not scratched when the block is free-standing. These are best adhered by applying a small drop of freshly mixed resin with the felt after the block has been polished. Figure 2:2 shows a jellyfish embedded in clear resin.

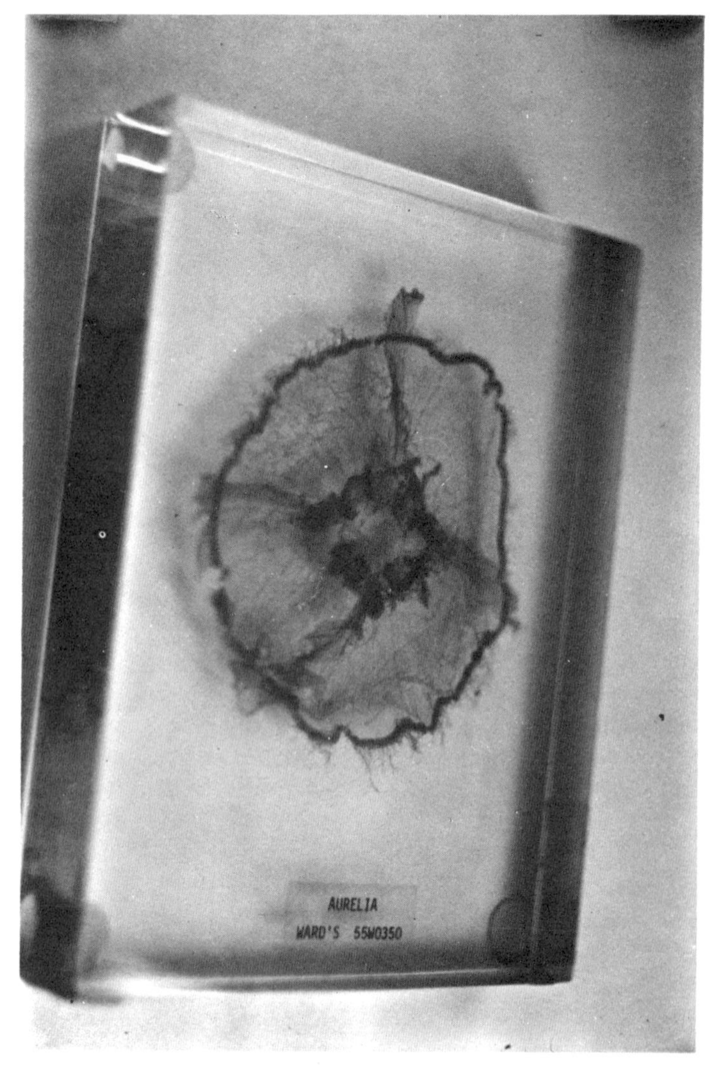

Fig. 2:2 Resin-embedded jellyfish (Gerrard & Haig)

Sources of materials

Resins, reagents and moulds

Ceemar Resins; E. M. Cromwell & Co. Ltd., Galloway Road, Rye Street, Bishop's Stortford, Herts.
Griffin & George Ltd., (*Kristablick* Embedding Kit), Ealing Road, Alperton, Wembley, Middlesex.
Trylon Ltd., Thrift Street, Wallaston, Northants.

Polishes

W. Canning & Co. Ltd., Great Hampton Street, Birmingham 18.

References

Rowland, J., *The Preservation of Specimens by Embedding in Cold Setting Resin*, E. M. Cromwell & Co. Ltd.
Trylon leaflets T 12 and T 78, Trylon Ltd.
Zechlin, K. (1971) *Setting in Clear Plastic*, Mills & Boon Ltd.

CHAPTER THREE

MICROSCOPIC WHOLE MOUNTS

MANY ANIMALS AND SOME PLANT MATERIALS ARE SUFFICIENTLY small to be best mounted whole on a microscope slide after suitable treatment. Related methods involved in histological preparations are dealt with in more detail in chapter nine.

Arthropod squash preparations

This technique is suitable for many small arthropods such as ants, small beetles, spiders, centipedes and millipedes. Some heavily pigmented specimens, especially beetles, may require bleaching beforehand. This may be done by treatment in a weak, say 1%, solution of potassium permanganate for about half an hour, followed by immersion in 5% oxalic acid.

Before the specimen can be squashed on a slide, it must be partially macerated by boiling in 10% potassium hydroxide. The time necessary for this depends on the specimen, but may be stopped when the hydroxide darkens. Most specimens are ready after about 30 minutes. After thorough washing with distilled water, the specimen is now placed on a clean glass microscope slide in such a position that when it is squashed the surface to be studied is uppermost and the appendages well displayed. Another slide is placed on top of the specimen and evenly pressed down, squashing the preparation. The two slides are firmly tied or clipped together and placed in 1% acetic acid for 5 or 6 minutes. It is then washed by immersion in distilled water and transferred through 50%, 70% and 90% alcohol, leaving for about 10 to 15 minutes in each. After careful removal of the top slide, the specimen should remain on the bottom slide, which is then flooded with two changes of absolute alcohol for 10 minutes each to complete dehydration.

After clearing for about 10 minutes in xylene, the specimen is covered by a suitable quantity of *Canada balsam* to support

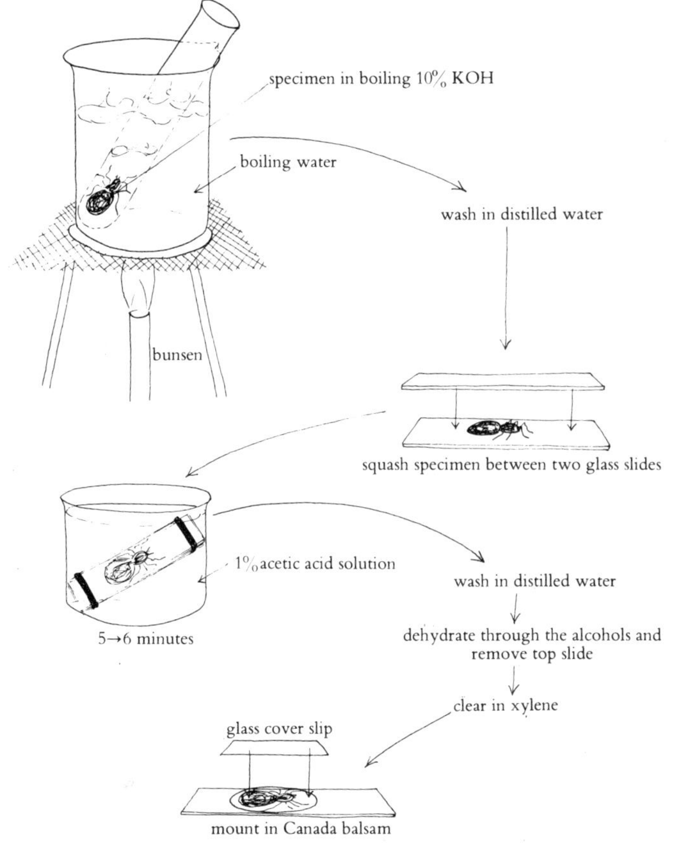

Fig. 3:1 Method for preparing an arthropod *squash mount*

a cover slip. Experience will show the amount of balsam to use, but it is better to use slightly too much rather than not enough. Any excess can easily be removed when it has set by wiping with xylene-soaked cotton wool. It is difficult to add balsam once the cover slip has been applied without trapping air bubbles or distorting the specimen. Figure 3:1 summarizes the whole procedure.

Once the mountant has dried (this may take a week or more depending on how much was used) the preparation is cleaned up and an appropriate label applied to the slide. (See figure 3:2.)

Chick embryos

Younger chick embryos, up to say 72 hours old, may be extracted from the egg and mounted whole for microscopic examination. Being delicate and small and surrounded by colloidal albumen, their removal undamaged from the egg requires great care. The following technique, first described to me by C. R. Pashley, has proved successful and with a little practice is fairly easy to perform.

Fertilized eggs can be purchased locally from a chicken farm. Place an egg horizontally on a suitable support, such as a ring of Plasticine, and carefully remove a portion of the shell from the top. This may be done using fine scissors, taking care not to cut too deeply where the embryo lies. Pipette off as much of the exposed albumen as possible, and if necessary gently roll the yolk with a camel-hair brush until the embryo is uppermost. Cut a paper ring just large enough to surround the embryo and any blood vessels and membranes required for the final preparation.

Place this over the *blastoderm,* gently pressing it with fine forceps. Now very carefully cut around the outer perimeter of the paper ring to a depth of about 3 mm into the yolk. The ring will support the embryo during this operation and

Fig. 3:2 Slide-mounted spider

enable it to be lifted clear of the yolk with fine forceps. Place the embryo in a dish of *Howard's ringer* solution and very gently move it backward and forward to clean off any yolk or albumen still adhering to the preparation. A camel-hair brush may be needed to remove any persistent debris. The formulation for Howard's ringer is

Sodium chloride	7·20 g
Potassium chloride	0·37 g
Anhydrous calcium chloride	0·17 g
Distilled water	1 litre

Now transfer the embryo to *Bouin's fixative* for about three hours. This contains

Saturated aqueous picric acid solution	75 cm^3
Formalin (40% formaldehyde)	25 cm^3
Glacial acetic acid	5 cm^3

Then transfer the embryo to 30% alcohol, replacing this with fresh alcohol of the same strength over a period of one hour. Now the embryo can be stained with *Borax Carmine* and then dehydrate by transfer through 30%, 50%, 70%, 90% and finally two changes of absolute alcohol, leaving for at least 15 minutes in each treatment. The paper ring support should be carefully cut free from the blastoderm and removed at the 70% alcohol stage. Take care that the blastoderm does not curl back on itself. After dehydration, the embryo is cleared by immersion in clove oil and finally mounted in Canada balsam beneath a cover slip. Figure 3:3 illustrates the techniques involved.

Small invertebrates

Small invertebrates such as some *Hydrozoan Coelenterates* and *Turbellarian Platyhelminths* are also usefully prepared as microscopic whole mounts. Specimens must be narcotized before fixation to ensure a relaxed posture. These animals invariably contract their whole bodies if the fixative is applied directly

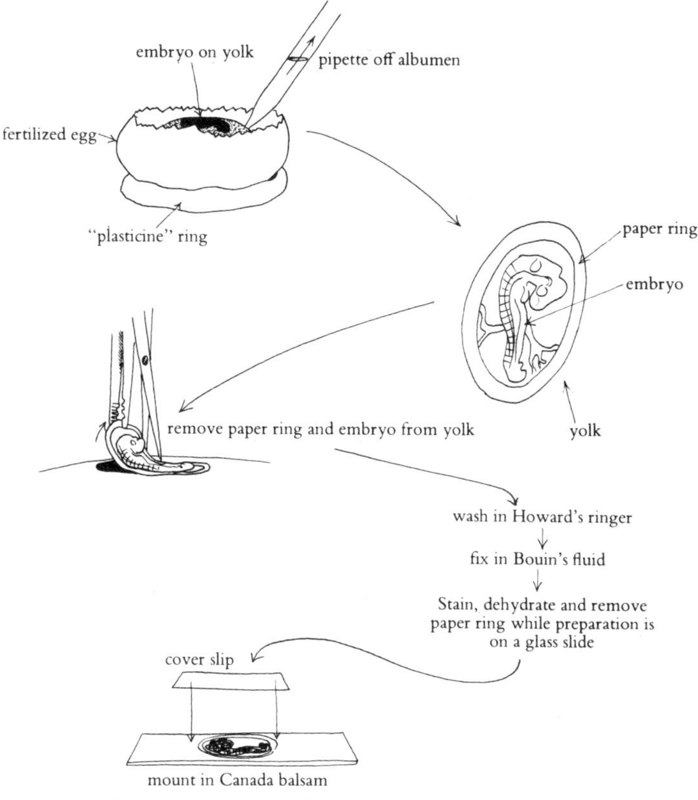

Fig. 3:3 Extracting and mounting a young chick embryo

to them. Many substances have been used for narcotization, but a weak, say 0·5%, *menthol* solution is good and easily prepared. Hydroid Coelenterates should be treated overnight. Specimens are then fixed in 70% alcohol for an hour or so. *Borax Carmine* may be used to stain any of these specimens, hydroids requiring treatment of ½ hour or more and small turbellarians, several minutes. Excess stain is removed by application of acid alcohol (1% hydrocloric acid in 70% alcohol). The extent of this differentation must be followed microscopically and stopped at the appropriate time by a wash with non-acidified 70% alcohol. Now the specimen is dehydrated, cleared and mounted in the way described in the previous sections of this chapter.

Many turbellarians are heavily pigmented and may be mounted unstained. Some preparations benefit from bleaching. This may be done by immersion in *sodium hypochlorite* after fixation in *Mayer's chloride* solution. This is made by adding a few drops of concentrated hydrochloric acid to a few crystals of potassium chlorate in a small dish. As soon as chlorine fumes appear, add about 50 cm^3 70% alcohol.

References

Mahoney, R. (1966) *Laboratory Techniques in Zoology*, Butterworths.

Phillips Dales, R. (1969) *Practical Invertebrate Zoology*, Sidgwick and Jackson.

CHAPTER FOUR

TAXIDERMY

TAXIDERMY IS AN OLD CRAFT AND AT PRESENT REPRESENTS THE best method for preparing most birds and mammals for display, while preserving their natural shape, texture and colouring. Taxidermy has been applied to fish, amphibia and reptiles, but with these, other techniques have proved more successful. Some of these are described in other chapters of this book. Considerable patience, experience and skill are required to obtain the best results, and professional taxidermists spend many years perfecting their methods. However, with a little determination and practice, very acceptable specimens may be prepared to meet the needs of the teaching situation or the school laboratory museum.

Measuring a specimen

In order to be able to arrange a completed specimen in a life-like position and to know how much stuffing material to put into a skin, it is essential that a series of accurate measurements be made before anything else is attempted. Where appropriate, these should include lengths of tail, legs, wings, ears, head and beak. Since it is into the thoracic and abdominal cavities that the stuffing will be inserted, measurements of the body are particularly important. Include the circumference of the chest and belly. It is often helpful to make a series of contact tracings of the specimen, as illustrated in figure 4:1. Other data that may be valuable should include notes on the

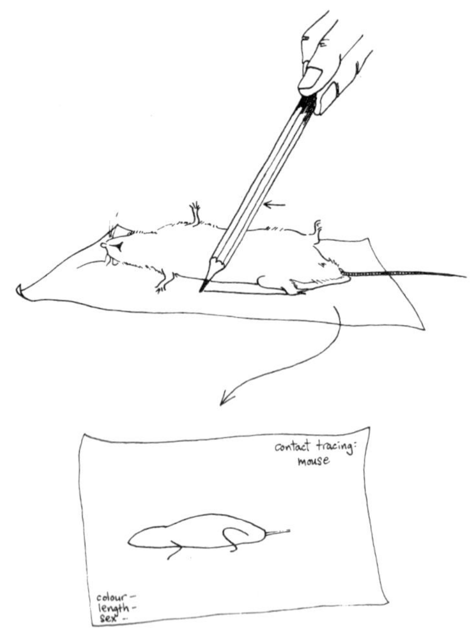

Fig. 4:1 Making a contact tracing

colours of all parts of the specimen, its weight, sex and where it was found. Trapping and killing animals for taxidermy is not to be encouraged. Specimens can often be found that have died from some natural cause or accident if one cares to look for them. They may be stored in deep-freeze if not required immediately.

Removing the skin

If the feathers or hair on a specimen are soiled with blood stains, it is best to clean these off before removing the skin. This is done by gently but firmly rubbing the dirty area with a rag or wad of cotton wool that is soaked with warm soapy water. Petroleum may be used to remove the more persistent blood or oil deposits.

To remove the skin from a specimen, it must first be cut. The object is to make an incision large enough for the body

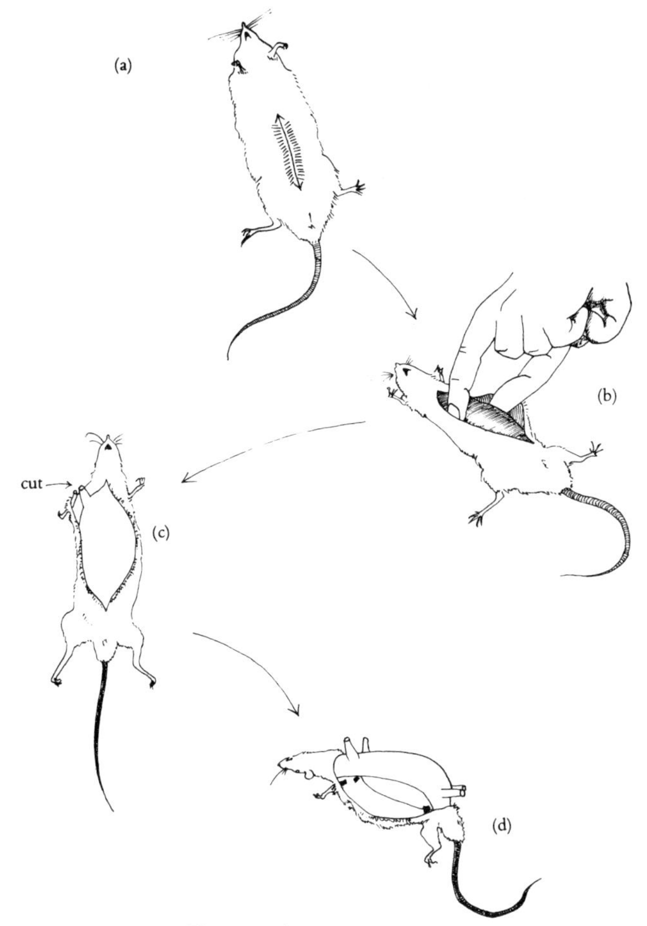

Fig. 4:2 Skinning a small mammal

to be pulled through, yet small enough not to be visible on the completed display. With most specimens, the best place for this incision is ventrally in the mid-line from the chest region down towards the genital area as illustrated in figure 4:2*a*.

Taking care not to increase the incision by ripping, the fingers may now be inserted into the body cavity and worked outwards along the inside of the skin, to free the body wall. This can be made easier by sprinkling borax powder onto the body and skin to dry the surfaces to be parted. Do this until the skin is separated from the body right up to the head, limbs and tail. (See figure 4:2*b*.) Cut the bones of each limb near to their points of articulation with the girdles.

Now turn the skin inside out as much as possible, bringing the freed body to the outside through the original incision. Next, free the head from its skin, taking special care with the ears and the eye-lids. Leave the skin still attached to the nose or beak. Remove the body by cutting the tail and the base of the neck. Make a contact tracing of the body and if necessary, in the case of birds, dissect it to determine the sex of the specimen before disposing of it.

Finally, push the upper portions of the limbs and a portion of the tail up to project from the inner surface of the skin. The bones of the skull, limbs and tail must now be thoroughly cleaned and all their flesh removed. This sequence of skinning is illustrated in figure 4:2.

Preserving the skin

The skin of birds and mammals is lined with layers of fat, particularly in aquatic types. This must be removed immediately but carefully with forceps and a small stiff-bristled brush before it decays. The choice of tools depends largely on the amount of fat to be removed and the strength of the skin. Always be cautious, yet thorough. Failure to remove all the subcutaneous fat may result in the feathers or fur falling off the skin later.

The skin must now be *cured* so as to preserve it from decay and to retain its supple nature. This may be done by rubbing liberal quantities of borax powder onto the inner surface of the skin in all the exposed regions. Skins of small mammals may benefit from a soaking for several days in carbolic acid. Alternative materials are available such as 'Lankroline', supplied by Watkins and Doncaster Ltd., 110 Park View Road, Welling, Kent. Its use is recommended after first fixing the skin in 5% formalin for several days, followed by thorough washing in water. The Lankroline treatment is then followed by rinsing in warm detergent (such as 'Teepol'), rinsing in water and pinning out to dry, fur or feathers uppermost.

Making and inserting a false body

Having made measurements and contact tracings of the carcass, it should now be possible to make an accurate replica of the removed body from some stuffing material. In the past, wadding, cotton wool, wood wool or gunmakers' tow have been used for this purpose. However, with medium or small-sized specimens such as squirrels or pigeons, it is much easier to sculpt a false body from a block of expanded polystyrene (figure 4:3*a*). Body contours can be far more accurately and quickly reproduced in this way, and the finished product will be lighter and easier to handle.

Having done this, the head, tail and limbs must be prepared. Wire is used to support these members and to fix them in position on the *body*. Soft iron wire is best as it can be bent easily, yet will support well. For the limbs use pieces sharpened at each end and about twice the length of the limb concerned. Push the wire beneath the skin through the palms of the hand or the sole of the foot, and then work it carefully up the limb

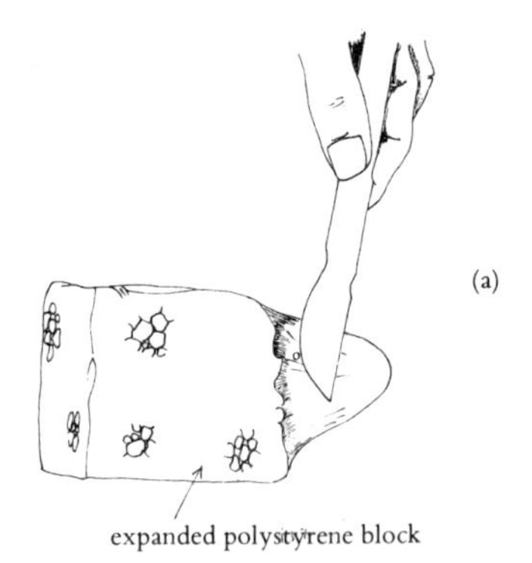

Fig. 4:3a Carving a false body from expanded polystyrene

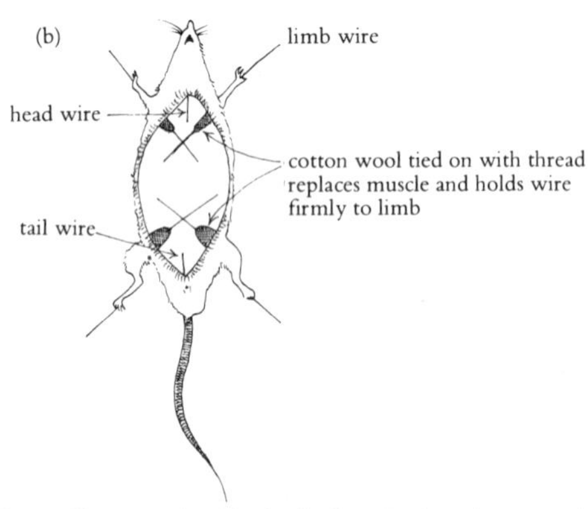

Fig. 4:3b Skin of a small mammal with the limbs wired ready to receive the false body

until it projects alongside the severed end of the bones. Be very careful at the joints or the wire may rip the skin. In bird wings, the wire may be pushed down the limb in the opposite direction. These wires and the ones from the head and tail (see figure 4:3*b*) will be used to fix the appendages firmly into

Plate 3 Pinned butterflies

the false body and allow them to be bent into a life-like posture.

The muscle on the skull, tail and upper limbs which was removed earlier, must be replaced with cotton wool wrapped around the exposed bone. This is particularly important on the limbs and face. The wrapping will also hold the supporting wires in place, and this may be effected by tying the cotton wool with fine thread or better, fuse wire.

These techniques are illustrated in figure 4:3.

Care must be taken to ensure that the limb wires are pushed into the body at the correct points and at the correct angles. A live animal or good photographs should be used as a guide, in conjunction with any contact tracing made of the specimen. Once this has been done, the skin may be wrapped around the limbs and *body* in its original position, and the incision made to remove it from the carcass must be sewn up. A fine thread, preferably of the same colour as the skin, should be used, but not so fine as to rip the skin when pulled tight. A fine semicircular needle will help in sewing up, reducing the need to lift the edge of the skin. The stitch used is a matter of personal preference, but it should be neat and fairly fine, so that it may be covered by fur or feathers on the skin. (See figure 4:4.)

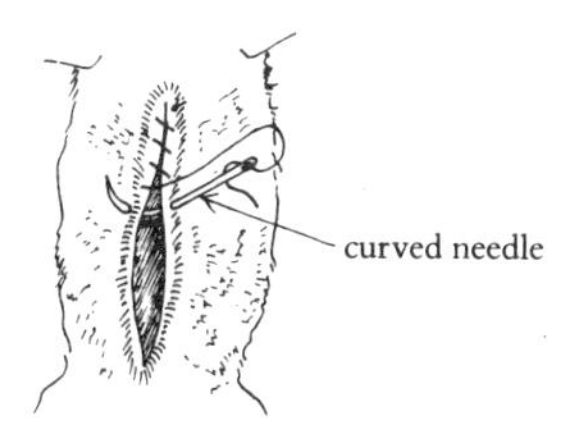

Fig. 4:4 Stitching up

Completing the specimens

The skin may now be *retexturized,* as it were, by shaking the whole preparation gently in a large box containing some *magnesium carbonate* powder. This should then be carefully brushed out of fur or blown out of feathers with a hair dryer. If a bird is being prepared, the feathers will have to be repositioned as in their natural state, for now, with sub-cutaneous muscle removed, they will tend to hang loosely. Start from the tail and work towards the head. The feathers can now be *set* in position by binding the whole preparation fairly tightly with thread, supported by several pieces of bent wire temporarily stuck into the false body through the skin in the mid-dorsal line. (See figure 4:5.) These materials can be removed after several days, leaving the plumage in the correct position.

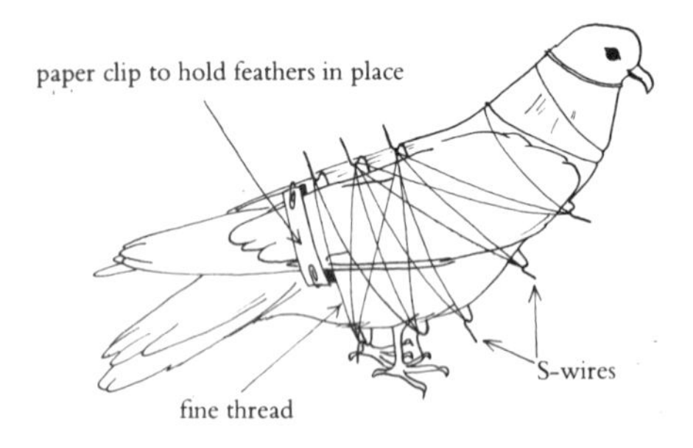

Fig. 4:5 Bird specimen bound with thread to set feathers in position

False eyes may be obtained from taxidermy suppliers such as Watkins and Doncaster Ltd., and are usually of foreign origin. One should specify the species for which the eyes are needed when ordering. The eye sockets must be cleaned out with a paste of borax and the glass eyes stuck into a base of modelling clay pushed firmly into the sockets.

The completed specimen may be mounted on a wooden or Perspex base, using the wires projecting from the limbs. Small items from the animal's habitat such as leaves and twigs always help to enhance the display. It will help in keeping a mount clean, to cover it with a glass or Perspex case, tightly fitted to the base.

Plate 4 (page 52) shows a specimen prepared professionally at Wollaton Hall Natural History Museum, Nottingham.

References

British Museum (Natural History), Instructions for Collectors No. 1 *Mammals* and No.2A *Birds*.

Pray, L. (1943) *Taxidermy,* Macmillan.

Wagstaffe, R. and Fidler, J. H. (1968), *The Preservation of Natural History Specimens* Volume 2. Vertebrates. Witherby.

CHAPTER FIVE

CASTING AND MODELLING

SOME SPECIMENS DO NOT READILY LEND THEMSELVES TO TAXI-dermy or other methods of dry preservation and display, in which case plaster-casting techniques can be employed effectively.

A plaster model, if well made, will show all the surface detail of a specimen and may be painted to give a life-like appearance. Once a good mould has been obtained, many plaster models can be made from the one specimen, enabling easy replacement of broken casts. Expensive or rare specimens such as fossils can be duplicated indefinitely for class use as well as display.

The primary mould

A primary mould giving excellent surface detail duplication can be obtained using a cold-setting moulding material such as 'Duplit' (produced by Amalgamated Dental Co. Ltd., London). Such materials are used commonly by dental technicians in the manufacture of dental appliances, and their quality is very high. Being *cold*-setting materials, the surfaces of delicate biological specimens such as frogs or toads are not harmfully affected. In fact moulds may be taken from anaesthetized specimens which can then recover later with no ill effects. However, the inexperienced are not advised to attempt this.

The specimen must be placed in as life-like a position as possible, and on a smooth solid base such as a glass plate.

Plasticine may be used to support parts of the specimen if necessary. It must be remembered that if the moulding material runs too far under the specimen, it will be impossible to remove the specimen when the mould has set. Such *undercutting* should be reduced to a minimum, although the finished mould will have some pliability. One ought to experiment to determine precisely how much undercutting can be allowed without damaging the mould. The prepared specimen must now be surrounded by a barrier to retain the wet moulding material as it sets. A cylinder of cardboard or even paper, provided the mould is not too large, will be sufficient for this purpose.

Now, when everything is ready for pouring, and not before, the moulding material is mixed. The precise mixing procedure may vary, dependent upon the type used, but this usually involves adding a measured amount of cold water to a given weight of the powder—twice the weight of water to that of the powder in the case of 'Duplit'. The compound must be thoroughly and quickly mixed so as to give a homogeneous paste. In this form it is now poured over the specimen, taking great care not to dislodge the specimen, nor to trap air beneath it. Any air bubbles produced in the pouring will rise to the surface of the mould before it sets. This can be aided if the whole procedure is performed on an electrically-operated bench vibrator. When this is done, the mould should be left to set for about 20 minutes, after which time the mould retainer is removed and the specimen carefully extracted from the mould. Figure 5:1 illustrates the sequence of operations described so far.

Duplit moulds begin to shrink on setting and should be used as soon as possible. If they must be kept for any length of time, they should be kept moist by wrapping them in soaked cloths or placing them in a damp chamber. However, there is no adequate substitute for immediate use.

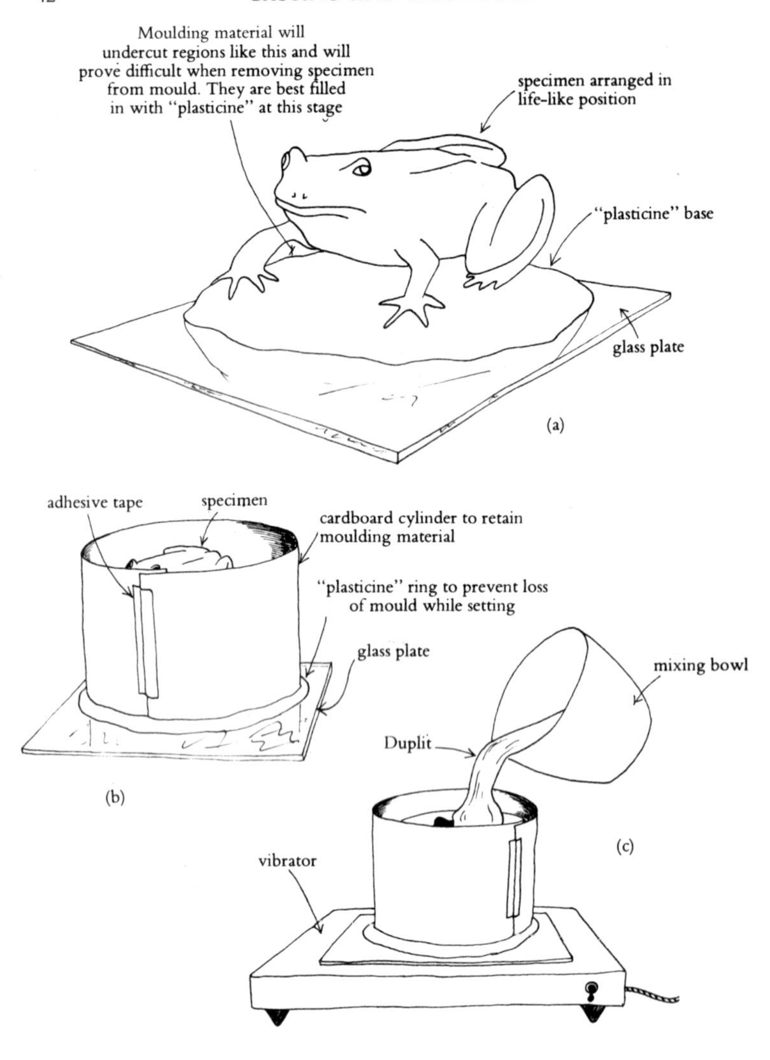

Fig. 5:1 Steps in the preparation of a primary mould of a frog

The primary casting

A primary mould may only be effectively used once and a primary casting is made from it using plaster of Paris. The chemical reactions involved in the setting process of Duplit create a slight acid exudation from the surface of the mould. If not dealt with, this causes the surface of a plaster cast to set more slowly than the bulk of the cast. To avoid the consequent loss of surface detail, the inner surface of the Duplit mould may be covered with a solution of potassium alum, usually provided with the moulding material. This is left for a minute or so, and is then poured out. Rinse the mould in water and pour the plaster casting as soon as possible. Mix the plaster of Paris quickly and thoroughly until a homogeneous fluid is obtained.

Pour this over the whole inner surface of the inverted Duplit mould to make sure air does not trap in any cavities it may contain.

Then fill the mould to the brim, gently prodding the plaster in the mould with a thin rod to help release any air bubbles. Again the use of a vibrator will help the plaster to settle, as well as make air bubbles rise.

The plaster must be allowed to set completely before it is removed from the mould. This setting process is exothermic and the cast is ready for removal when totally cold. Retrieval of the plaster cast involves sacrifice of the primary mould which will eventually shrink and become brittle. It is best to break the mould away from the cast so as not to damage the plaster.

The secondary mould

If more than one casting is required, it will be necessary to prepare a secondary mould of a more permanent nature. This may be made from rubber latex such as 'Vinamould' (this is produced by Vinatex Ltd. of Carshalton, Surrey, and may be obtained from Alec Tiranti Ltd., 72 Charlotte Street, London W1P 2AJ). Rubber latex moulds give good surface detail; they are tough yet pliable, enabling fairly complex casts to be removed from them without damage to the mould or the cast. These moulds can be stored and used indefinitely and may even be cut up, melted down and used again.

Vinamould hot-melt compound melts at between 120° and 130° C and must be heated to that temperature before it can be poured—hence the necessity to use a cold-setting moulding material on the original specimen. Cut a slab of rubber latex into small cubes with edges of about 1 cm. Heat these in a pan kept specially for this purpose over a moderate flame and in a fume cupboard, or at least by an open window—the fumes produced are somewhat uncomfortable. Stir the latex as it starts to melt. Half a kilogram of Vinamould will melt in about half an hour.

Because of the porous nature of its surface, a plaster cast will have to be sealed before pouring molten latex over it. This prevents the mould sticking to the cast while setting. A coating of varnish may be used for this, applied with a fine paint brush as thinly as possible so as not to even out any surface features on the cast. Varnish S.P.99 (also supplied by Alec Tiranti Ltd.) has proved effective for this purpose.

Surround the sealed cast with a metal mould-retainer (an old clean food can is adequate for smaller pourings) pushed into a Plasticine ring to prevent loss of latex at the base. Pour the latex around the sides of the cast, but never directly onto it. The heat of the latex rising up the side of the cast should drive off any pockets of air. (See figure 5:2.) Be sure to pour sufficient latex to more than cover the cast, so that when it is removed no part of the mould is too thin. Allow to set overnight before removing the cast. Plates 5 and 6 illustrate plaster casts of a fish and a frog (Wollaton Hall Museum, Nottingham).

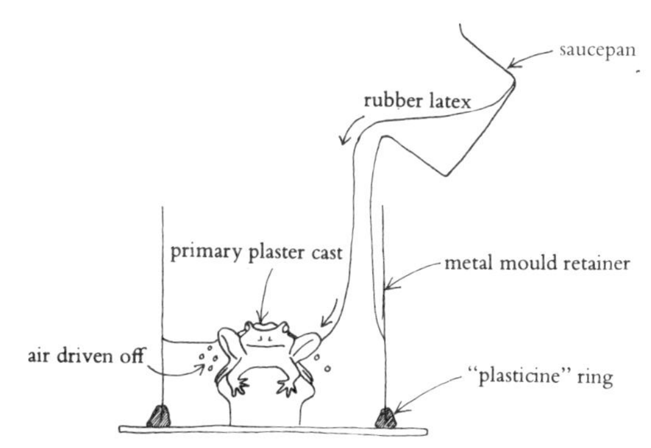

Fig. 5:2 Preparing a permanent latex mould from a primary plaster cast

Plaster casts can be painted to give life-like appearance although this involves considerably more skill and patience than the preparation of the casts themselves. Casts of fossils can be given a very effective 'rock-like' appearance by soaking them in a saturated solution of potassium permanganate.

The need to produce a primary mould and cast, and subsequently a secondary mould, can be overcome by using a cold-setting moulding material that has the lasting and durable qualities of rubber latex, without the need to use high temperatures. Cold-cure Silastomer 9161 (plus catalyst N9162) may be used here, although it is more expensive to buy. It is produced by the Midland Silicon Co. Ltd., Glamorgan.

Plant material is notoriously difficult to preserve and display. Moulding techniques, however, may be used in some cases. Moulds of individual leaves or petals may be obtained as follows.

Lay the leaves flat on modelling clay and gently press their edges down to eliminate undercutting. Then cover with plaster and allow it to set. When the clay and leaves are removed, wax casts may be obtained from the impressions left in the plaster. A 50:50 mixture of beeswax and paraffin wax will give good results. While this mixture is molten, the addition of a little Canada balsam will make it harden on setting. Take a piece of crepe paper a little larger than the leaf to be cast, dip it into the molten wax mixture for a few seconds, allow excess to drip off, and then press it firmly into all the

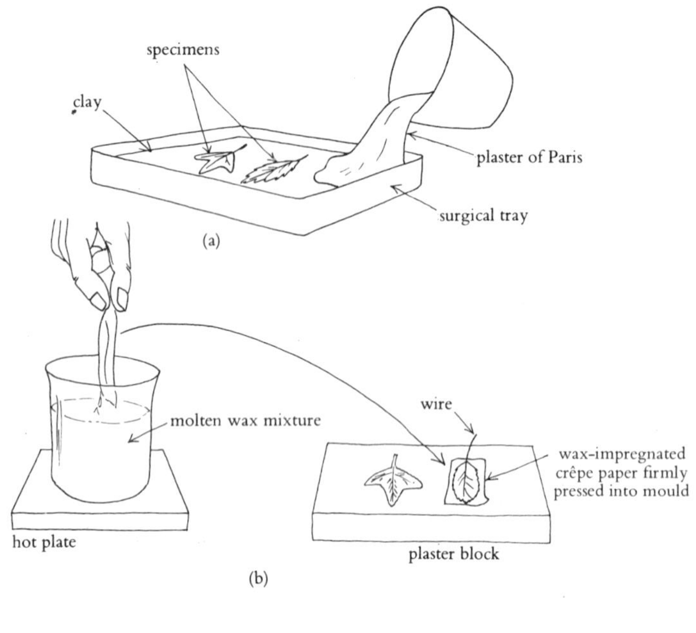

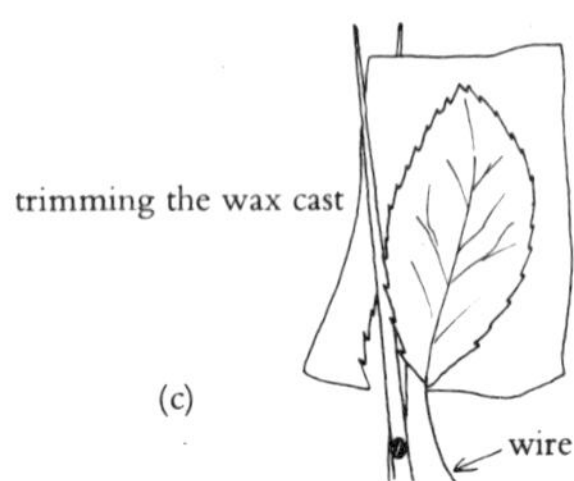

Fig. 5:3 Method for producing leaf casts

depressions in the mould. A piece of thread-covered wire is now laid along the central vein of the 'leaf' and a second piece of wax-impregnated crepe paper is applied as before. After setting, it is removed from the plaster mould and the edges trimmed to the correct shape with scissors. The use of green crepe paper in this preparation minimizes the need to paint the cast on completion. Figure 5:3 summarizes this technique.

Gratitude for help in the preparation of this chapter is expressed to Mr. R. Breen, lecturer in dental technology at People's College of Further Education, and also to Mr. Sharp, Wollaton Hall Museum, Nottingham.

SECTION B

PREPARATION OF PARTS OF SPECIMENS

CHAPTER SIX

DISPLAY DISSECTIONS

DEMONSTRATION OR DISPLAY DISSECTIONS ARE USEFUL IN TEACHing the internal anatomy of a variety of animals. Pupils who are to dissect specimens themselves may make a more meaningful attempt at this after first studying such an exhibit. Techniques of dissecting particular anatomical systems in specific animals are beyond the scope of this book and are adequately described and illustrated by Rowett (see references). Here follows an account of some of the methods that may be used to inject blood vessels *in situ* with coloured media. These will highlight the vessels and greatly enhance their display.

Many media or injection masses have been employed in this work, but the best results are obtained with coloured *rubber latex*. Several types are available and may be purchased from most biological suppliers. The author has used latex injection mass supplied by Gerrard and Haig Ltd., Worthing Road, East Preston, Sussex.

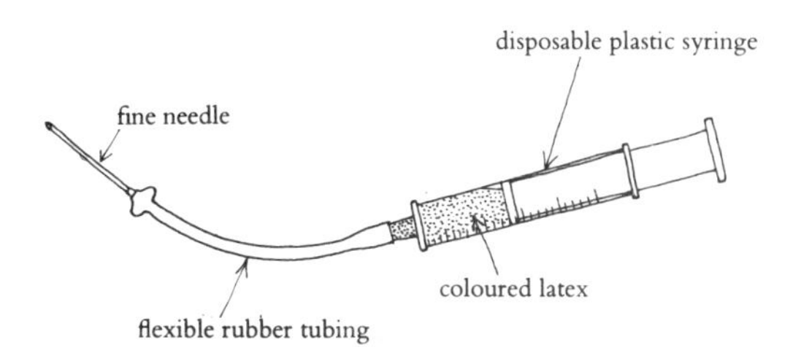

Fig. 6:1 A syringe with a movable needle for latex injection

The method used to make the injection depends largely upon the size of the appropriate vessel. Often a simple hypodermic syringe fitted with a fine enough needle is best, being easily handled and filled with latex. If the needle is separated from the syringe by a short length of flexible tubing, then the needle may be more easily manipulated into the injection site. (See figure 6:1.) For even finer work, say when injecting small invertebrates, glass micro-pipettes should be used. These are fairly easily made from small-diameter soda-glass tubing that is drawn out from a central point where it is heated. Provided the right pull is applied, with a little practice one piece of such tubing will produce two fine-tipped micro-pipettes. (See figure 6:2.) Untidy ends are easily cut off when the glass is cool.

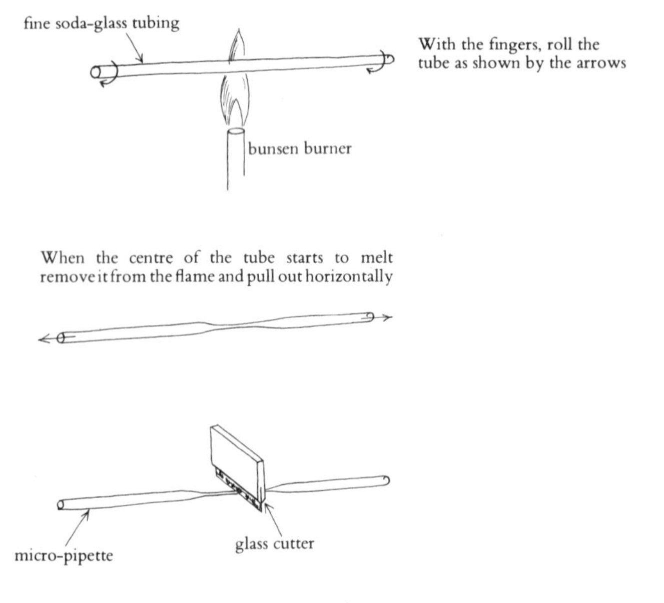

Fig. 6:2 Method for making micro-pipettes

Plate 4 A stuffed *Jay*

If large volumes of latex are to be injected and require a constant moderate temperature, a simple apparatus may be used, such as that illustrated by figure 6:3.

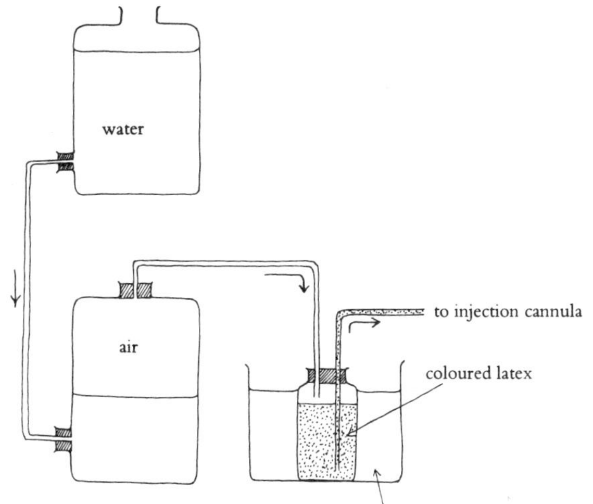

Fig. 6:3 Apparatus for latex injection

When introducing a micro-pipette or a needle into a vessel for injection, great care must be taken not to lacerate the vessel so much that the latex leaks from the injection site. For this reason it is advisable not to use a cannula which has a sharp end. Incision is made in the vessel with fine scissors or a scalpel before introducing the cannula. Suitable thread ligatures should be used to hold the cannula in place while injecting. (See figure 6:4.)

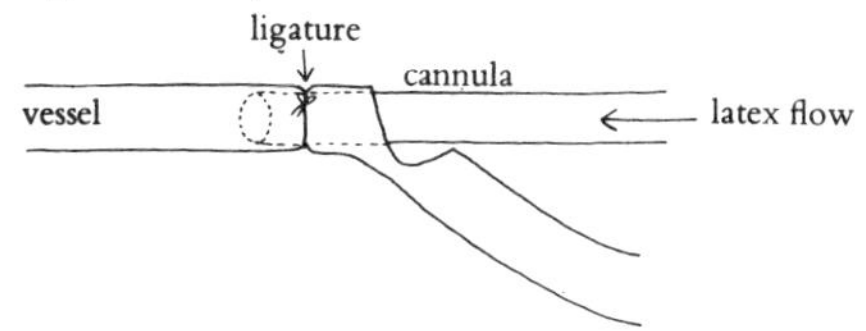

Fig. 6:4 Ligated vessel with cannula inserted

Injection methods for particular animals

Only injection methods for the more common laboratory animals will be described, but injection of other animals should be successful with a little care and common sense. There is considerable room for experiment. In any case, injection should not be delayed after the death of the specimen, and often it is useful to *flush* the blood vessels first with warm normal saline. (Mahoney recommends a 0·25% solution of sodium nitrite to dilate the vessels.)

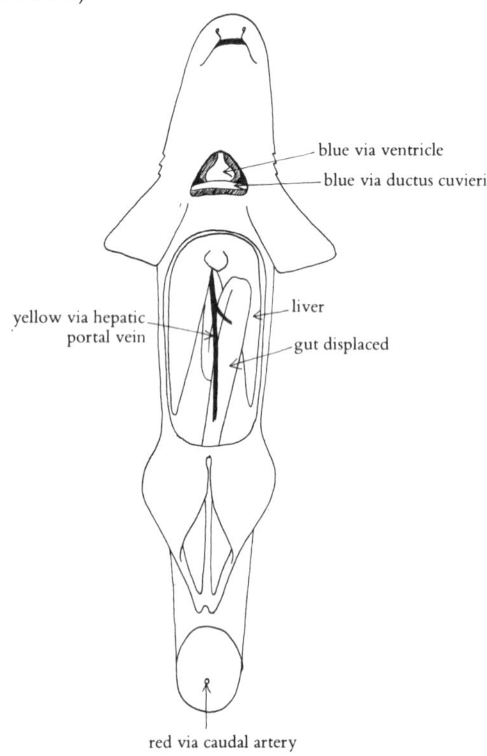

Fig. 6:5 Sites of injection for dogfish

Plate 5 A painted plaster-cast of a *tench*

a. Dogfish

Fresh specimens are essential for good results. The afferent branchial arteries to the gills are injected, conventionally with blue mass. The injection is made in a forward direction through the ventricle or *conus arteriosus* of the heart. The rest of the arterial systems is injected with red mass, through the *caudal artery*, revealed by removing the tail.

The venous system of cartilaginous fish consists of many fairly large sinuses and consequently requires a relatively large volume of injection mass, usually blue. This may be introduced via the *sinus venosus* of the heart or the *Cuvierian ducts* laterally. The portal veins draining blood from the gut to the liver may be injected with a yellow mass. Injection is made through the *hepatic portal vein* away from the liver. Figure 6:5 illustrates these injection routes.

b. Frog

The frog is probably the best animal with which to gain initial injection experience. It is easy to obtain freshly-killed specimens

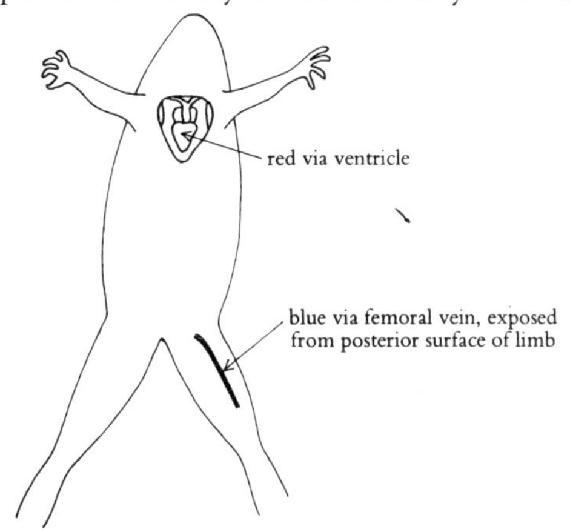

Fig. 6:6 Sites of injection for frog

and requires only a few cubic centimetres of injection mass. Inject the arterial system red via the ventricle. The venous system may be injected (blue) either via the sinus venosus of the heart or one of the *femoral veins* in the hind limbs. Do not rush the injections. Only gentle pressure is necessary and good results are obtained with a hypodermic syringe. (See figure 6:6.)

c. Rabbit

Immediately the animal is dead, sever one of the vessels in the hind limb and drain the blood from the entire vascular system. Then flush the system with warm saline through the sites to be used for latex injection later. First inject the hepatic portal vessels with a yellow mass, injecting through the portal vein near to the liver, towards the gut. It will be necessary

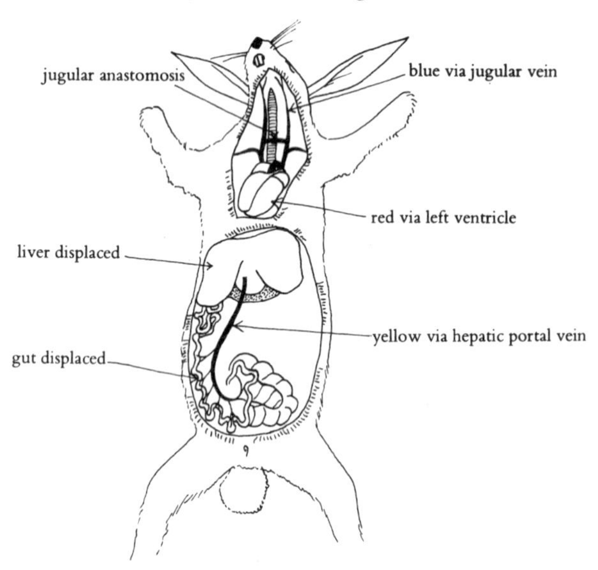

Fig. 6:7 Sites of injection for rabbit

to manipulate the gut with the fingers to allow complete injection.

Next inject the venous system with a blue mass. This may be done via the sinus venosus of the heart or, better, via the right *jugular vein*. The injection is made towards the head and the mass will cross via the *jugular anastomosis* into the rest of the venous system.

Finally, inject the arterial system with a red mass via the left ventricle. Figure 6:7 illustrates these injection routes.

With care and practice, many excellent injections can be made. The animal may then be further dissected to display the vascular system and its supply to the organs of the body clearly. The completed dissection may be permanently mounted in fixative in a museum jar such as that shown in plate 7 (page 69).

References

Green, T. L. (1936) *Zoological Technique*, Allman.

Mahoney, R. (1966) *Laboratory Techniques in Zoology*, Butterworths.

Rowett, H. G. Q. (1962) *Guide to Dissection*, John Murray.

CHAPTER SEVEN

CORROSION CASTING

THE BLOOD SUPPLY TO THE ORGANS IN AN ANIMAL'S BODY IS VERY complex. Direct dissection of an organ such as the lung, brain or liver will reveal only the larger blood vessels. The intricacy of the great mass of vessels, down approximately nearly to the terminal arterioles, can only be fully appreciated and studied if a technique such as corrosion casting is employed. Basically the method involves injecting a coloured resin into blood vessels, air passages, pelvis of kidney, etc., and then corroding away the tissue to leave a resin cast of those vessels. D. H. Tompsett has extensive experience of these techniques, and the following descriptions are largely based on his writings. The author is grateful to Dr. Tompsett for his advice in preparing this chapter. Below is given in outline the method followed for treating a typical organ, although some organs require special preparation outside the scope of this book.

Preparing the organ

The organ must be removed from a freshly killed animal and not damaged in any way. Immediately after removal, the organ is either stored in a polythene bag in deep-freeze at a temperature of -20° C or lower, or the blood must be flushed out and the tissue fixed to prevent blood clots and decomposition. This may be done by perfusing the organ with weak formalin solution, usually 3% made up in deaerated water and leaving in this overnight; longer would cause excessive hardening.

Waterlogging occurs during fixation of some organs which have to be squeezed firmly with the hands to displace enough water to allow the resin to flow easily into the vessels. In other organs such as with portal or hepatic veins, even without squeezing, injection of resin must be limited to prevent excessive penetration into the tissues.

After cannulation, fixation is performed with the organ submerged in water. Before fixation, air, which is invariably present in the vessels of isolated organs, is dislodged by gentle squeezing of the organ under water, until no more bubbles escape from the cannula. The resin injection is also done under water which supports most of the weight of the organ, avoiding distortion. Tepid as opposed to cold water is used for most organs.

Injecting the resin

Unsaturated polyester resins, similar to those used for embedding (see chapter two) are injected into the vessels to be cast. Several of these resins are available from different manufacturers. The following formulations are based on those supplied by Trylon Ltd., Wollaston, Northants. The standard injection mass recommended by Dr. Tompsett is made up as follows:

Trylon resin	100 g
Trylon thinner	20 g
Catalyst paste	12 g
Trylon activator	6 cm^3

Any of a wide variety of pigments may be added to this mixture, enabling different vessels in an organ to be injected with different colours. Conventionally, arteries are injected red and veins blue. Other cavities such as the pelvis and calices of the kidney or gall bladder and bile ducts of the liver may be yellow. (See figure 7:1.)

On addition of the activator, the resin mixture rises in

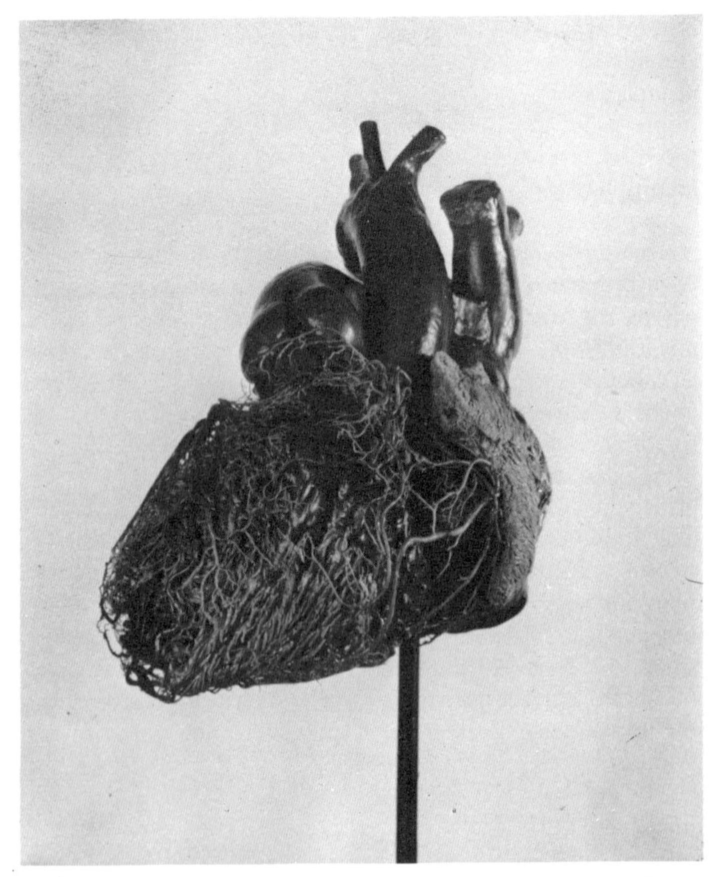

Fig. 7:1 A resin cast of the blood vessels of a human heart including the pulmonary trunk. This illustration was made by Dr. D. H. Tompsett from a cast in the Wellcome Museum of Anatomy and Physiology in the Royal College of Surgeons of England.

temperature, setting at about 35° C. It is recommended that the mixture be adjusted to about 20° C before adding the activator, and that the temperature be continuously monitored. If the resin is injected when at about 27° C, there will be around four minutes for this procedure to be completed. If injection is made too early, the organ may be over-injected, and finer branches of the cast remain permanently tacky. Resin may also impregnate the walls of vessels. If it is left too late, it may be too thick to flow adequately in the vessels and the cast will be incomplete. Obviously several attempts at the technique must be expected before satisfactory casts are obtained. Experience is necessary to determine the particular problems of injection with different organs. The size and complexity of the vessel network being injected will also govern the detail of the injection procedure.

When the resin is injected, it forces the fixative remaining in the vessels out into the surrounding tissues. Hence the need for these not to be waterlogged. Some fixed tissues, such as those of the lung and liver, are readily permeable to water and present no problem in this respect. Other more dense tissues like the kidney may require further treatment to enable adequate resin injection. Tompsett suggests injecting the veins first, the water in these being displaced via the arteries and the pelvis. Then inject the arteries. These have a relatively small total volume and little water is driven out. Finally, resin is injected into the pelvis, having first made a small puncture in the upper wall of this cavity to allow displaced water to escape without interfering with the injection.

The injection apparatus devised by Tompsett is simple, easily made and highly effective. It is shown in figure 7:2. The sphygmomanometer bulb enables the resin to be forced into the vessels of the organ at the pressure necessary to give a good injection. This will be determined by experience.

Tissue corrosion

To ensure that the resin is completely cured before the surrounding tissue is corroded, the injected organ should be left to stand in water for eight days. It is then transferred to a bath of

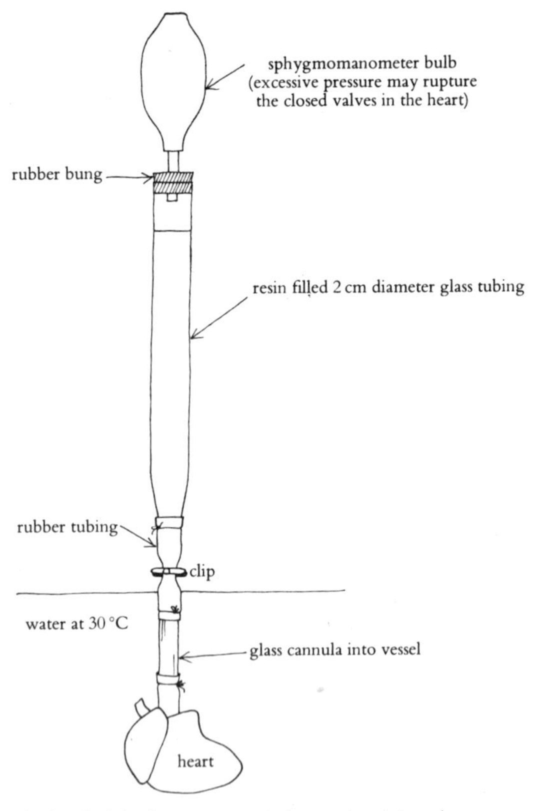

Fig. 7:2 Standard resin injection apparatus being used to inject the coronary arteries of a heart fixed with aortic valves closed

concentrated hydrochloric acid (specific gravity 1·16 to 1·18). All due precautions should be taken at this stage, including the use of heavy rubber gloves, protective spectacles and a carrying tray for the injected organ to avoid injury in case of an accident. Figure 7:3 shows a suitable arrangement which should be kept in a fume cupboard. The process of corrosion should be continued for the minimum time necessary, as prolonged soaking

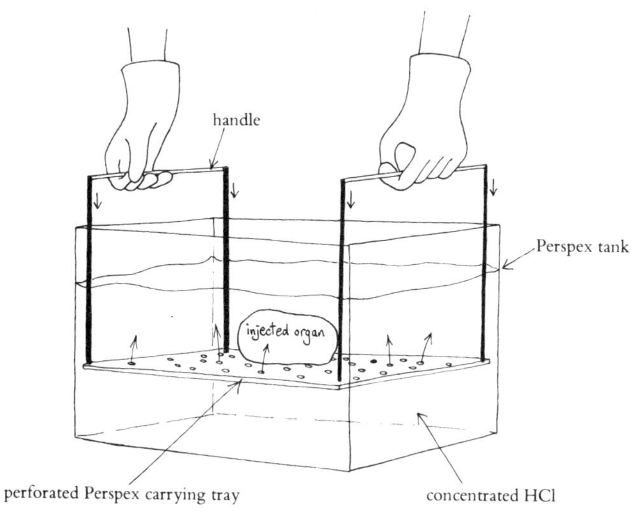

Fig. 7:3 Immersing an injected organ in concentrated hydrochloric acid

in acid attacks some pigments. This may be from two days to two weeks depending upon the strength of the acid and the size of the organ.

Preparing the cast for display

When corrosion is complete, the remaining resin cast will be much more fragile without its tissue support. Consequently great care has to be exercised in handling the cast which must now be thoroughly washed in tap water. This is done by directing a fine jet from a large hypodermic needle attached to the tap by hard polythene tubing. The needle is held just above the cast so that a mixture of air and water is directed onto the cast. The pressure used should be sufficient to remove remains of macerated tissue, but not so great that the cast breaks. Any relatively large fragments of the cast that may be broken off may be stuck back on later with resin cement, or retained and used to repair other casts that may be damaged at a later date. Any unwanted resin, such as may appear if some vessels were damaged during injection, allowing the resin to impregnate the tissues, may be removed with forceps or sharp cutters. Care must be taken not to damage the rest of the cast during such pruning. In most cases this hazard is very greatly reduced by first soaking the cast in warm or hot water to increase the flexibility of the resin.

The appearance of finished casts may be improved by spraying them with a resin which sets with a non-tacky surface, such as preactivated Trylon Resin CL 202 PA (diluted with acetone) to which has been added 4% catalyst paste just before use. The acetone thins the resin and allows application from an atomizer. It must be evaporated as the resin is sprayed on, by blowing warm air onto the cast; otherwise the cooling due to evaporation of the acetone may cause moisture in the air to condense on the cast and make the resin spray milky in colour. A domestic hair drier works well.

Robust casts may be mounted in Perspex display cases, but more delicate casts are better embedded in clear resin (see chapter two). In this form they are easily handled and stored.

It must be emphasized that this brief account is only a general outline of the methods involved in corrosion casting and is intended to stimulate interest in this work. Details of more specific methods for particular organs may be obtained from the references below.

References

Tompsett, D. H. (1970) *Anatomical Techniques*, 2nd. edition, Livingstone.

Trylon leaflet T.26.

CHAPTER EIGHT

SKELETAL PREPARATIONS
(including transparencies)

DISPLAY PREPARATIONS OF SKELETONS MAY BE MADE EITHER BY removing the flesh and presenting the skeleton on its own or, as is often the case with small and delicate specimens, the skeleton is stained to show through the flesh around it, which is made transparent. Particularly with the former method it is vital that a thorough knowledge of the skeleton be acquired before attempting the preparation. This will help prevent damage or loss and will pay dividends when the cleaned components of the skeleton are reassembled.

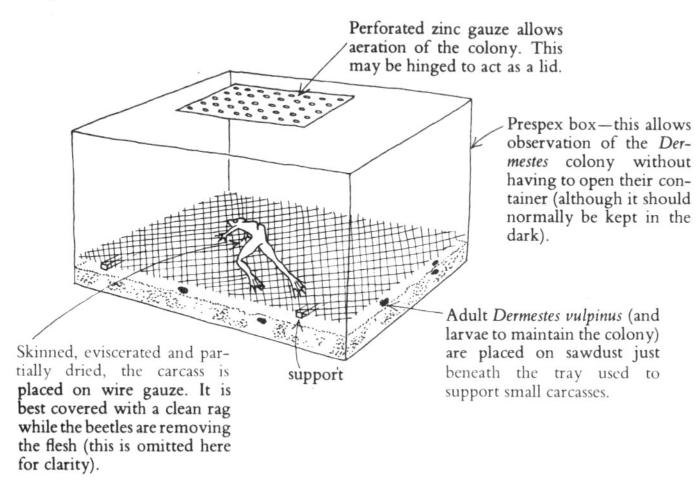

Fig. 8:1 A simple *Dermestarium*

The removal of flesh from a bony skeleton

The flesh may be removed in a number of ways. If the specimen is freshly dead and not too big, it may be presented to a colony of leather beetles *Dermestes vulpinus* which will clean the bones very well, saving the technician much time and effort. A suitable arrangement for the maintenance of such a colony is shown in figure 8:1.

Similarly a carcass may be buried in a suitable place and disinterred after time enough for soil animals, fungi and bacteria to remove the flesh. Certainly very clean undamaged skeletons may be prepared in this way, but the time involved is very long and there is real danger of losing small bones.

Quicker methods involve the maceration of the tissue from the carcass. Whichever maceration technique is used, first remove as much flesh as possible using plastic forceps—metal instruments may scratch bone.

a. Hot water maceration involves subjecting specimens to a very hot but not boiling, weak solution of sodium carbonate. The specimen is taken out of this from time to time and loosened flesh is removed with forceps. Extreme care must be exercised here as it is very easy to lose small bones among the tissue débris. Any tissue still clinging to the bones after this treatment must be removed with the aid of a small brush (a toothbrush is a good instrument). Application of a paste of calcium hypochlorite at the same time will aid the removal of the last of the flesh as well as initiate the process of bleaching.

b. Enzyme maceration involves the use of proteolytic enzymes such as the pancreatic digestive enzyme *trypsin*. This will hydrolyse the protein constituents of muscle and other tissue, releasing them from the bone to which they are attached. This hydrolysis is promoted by incubation at about 37 °C (body temperature of many mammals). A 1% solution of the

Plate 6 A painted plaster-cast of a frog

enzyme is used in a weak solution (say 0·5%) of sodium carbonate which provides a pH in which the enzyme can operate. The use of a fume cupboard is strongly advised for the incubation which should be left for 1 to 2 days (the smell of digesting carcasses will not be appreciated).

Bones cleaned by either of these maceration techniques are thoroughly washed in running water to remove all traces of the materials used and any persistent débris. If necessary the bones may now be bleached in a 50% solution of hydrogen peroxide containing a few drops of ·880 ammonia.

Because of the dangerous nature of the chemicals used to defat large bones (e.g. immersion in trichloroethylene) it is advised that only relatively small specimens be prepared. The sodium carbonate used in each maceration process will remove most fat from these.

These methods are summarized by the sequence of illustrations shown in figure 8:2. Following bleaching, the bones are dried in an incubator and are then ready for mounting.

Mounting skeletons

The method used to mount cleaned bleached bones for display depends largely on the function they have to perform. Skulls, vertebrae, or bones of the limb and girdles may be kept separately in boxes or mounted with glue on black card. In this way they may be stored and studied individually and with little fear of damage. However, much more valuable information about the overall functions of a skeleton in the body can be obtained from a fully articulated mount. The bones are joined in the positions they normally occupy in relation to one another in the living animal. An articulated preparation is much more fragile and difficult to store and use without damage than individual bones, especially in the teaching situation. When preparing an articulated mount, a compromise has to be achieved between ensuring the strength and durability

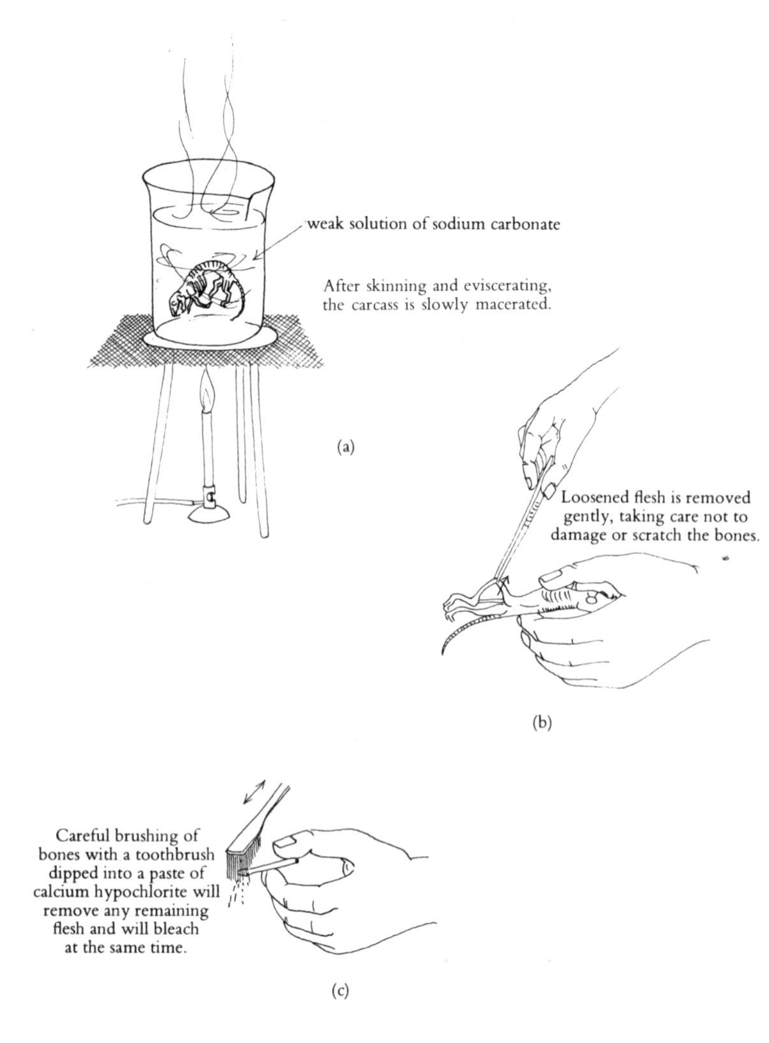

Fig. 8:2 Steps in the removal, cleaning and bleaching of bones from a small carcass by *hot-water maceration*

Gerrard & Haig

Plate 7 The bronchial arteries of a dogfish injected with coloured latex

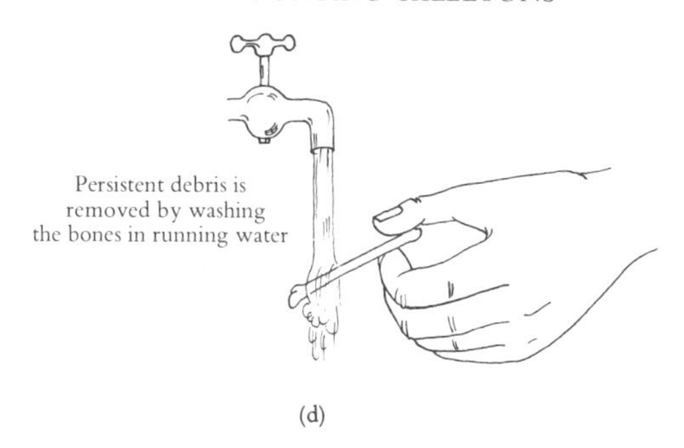

(d)

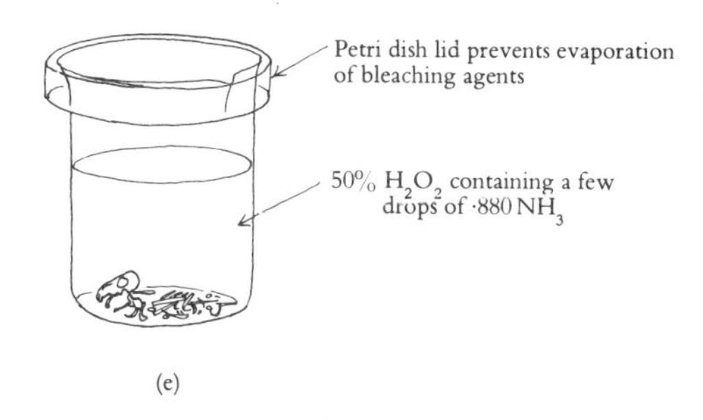

(e)

of a specimen and destroying its appearance by excessive use of glue or wire. Effort to retain ligaments between bones during maceration will minimize reassembly problems later.

Small specimens such as mouse or frog skeletons may be articulated entirely with the use of a transparent glue (*Airfix* is admirable). Great care and skill is needed to achieve a 'life-like' position, and it is a good idea to use a photograph or drawing of the living animal as a guide. Start by gluing the vertebrae together, making sure to arrange them in the correct sequence. (This is made easier if care is taken during maceration to keep the vertebrae in order.) A length of fine wire threaded through the neural canals of the vertebrae will add strength to the preparation, and the vertebral column can then be bent into the required shape, and the bones glued afterwards. To avoid the use of excess glue, apply the adhesive with a pin-head or touch the bone's surface onto a blob of glue rather than trying to squeeze the glue from its tube directly onto the bone. At the anterior end, the wire used to support the vertebral column can be pushed into adhesive in the cranial cavity of the skull, fixing this element firmly in the required position.

When all the adhesive is dry, glue the pectoral and, if separate from the vertebrae, pelvic girdles into their appropriate places, bearing in mind the eventual positions of the limbs. It may be necessary to support the skull and vertebrae while the glue is setting. Then, again using appropriate supports until set, attach the bones of the limbs one at a time. It is convenient with small specimens to fix the bones of the hands and feet to the base with glue and then to fix the limbs to these. The preparation must be held in place above the base while this is done to make sure the limbs meet up with the hands and feet in the correct place. Figure 8:3 summarizes these procedures.

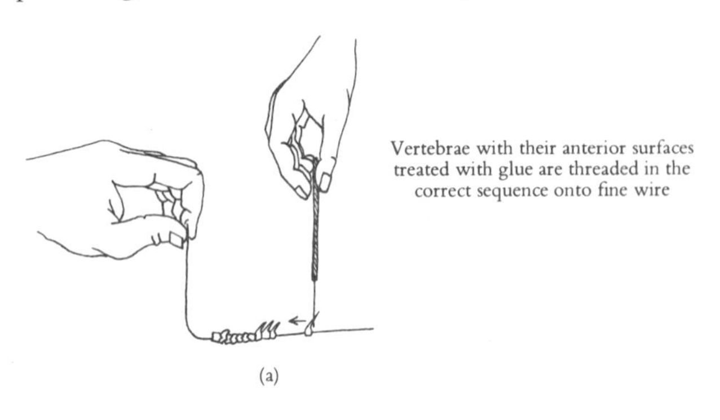

Fig. 8:3 Mounting a small bony skeleton

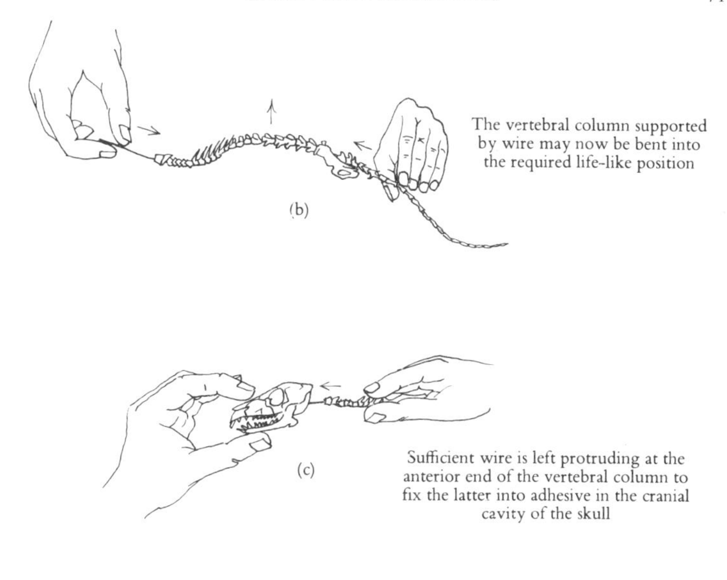

The vertebral column supported by wire may now be bent into the required life-like position

Sufficient wire is left protruding at the anterior end of the vertebral column to fix the latter into adhesive in the cranial cavity of the skull

The skeleton, as it is glued together a piece at a time, must be supported (matchboxes are useful) in its final position. Transparent Perspex rods for permanent support on the base board may be fixed more securely and left in position.

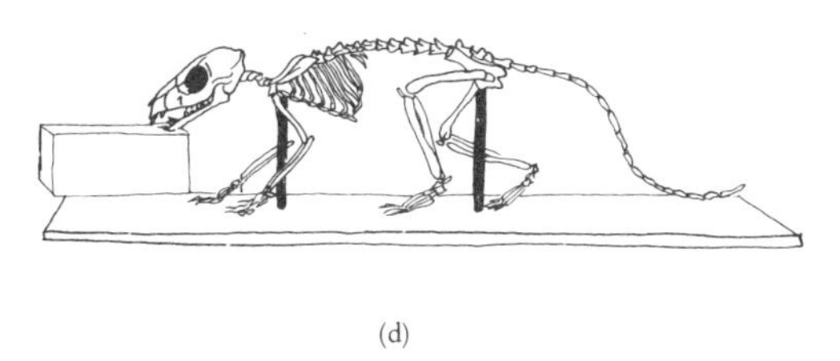

(d)

The articulation of larger bones may require wiring to give the strength needed to support them. This often also allows the bones to move against one another in the way they would in the living animal. Figure 8:4 illustrates a mounted dry skeletal preparation.

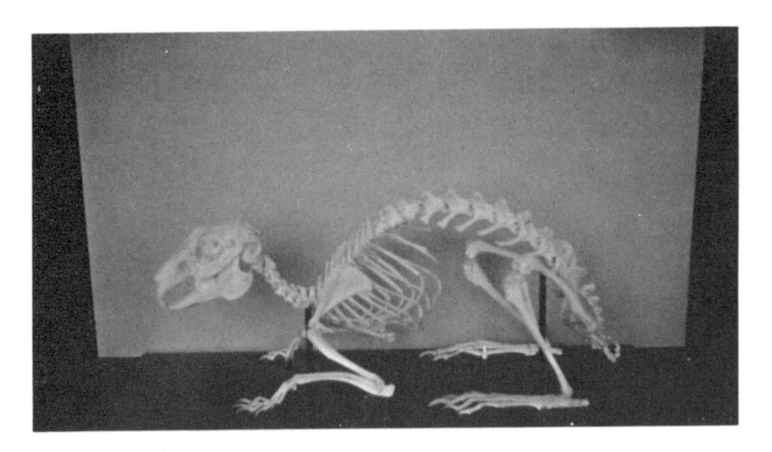

Fig. 8:4 The mounted skeleton of a rabbit (Gerrard & Haig)

TRANSPARENCY PREPARATIONS

Very small and delicate specimens, particularly those which are not fully formed, such as embryos, may be displayed as part of a *transparency preparation*. The specimen is first subjected to a stain which is taken up by the skeleton. Then after suitable treatment it is mounted in a clearing agent which renders the tissues transparent, the stained skeleton showing through. Two dyes are in common use here: *Victoria Blue* which stains cartilage and *Alizarin Red* which stains bone. Excepting embryos which may be treated in the entire state, it is advisable at least to skin and eviscerate specimens prior to treatment. This is especially necessary for animals with tough or pigmented skins which will not completely clear. Excess muscle may also be removed from larger specimens. In any case the specimens should be neatly treated, leaving no ragged surfaces.

a. Victoria Blue

Specimens should be fixed in 5% formalin for a few days prior to treatment. Then, the first step is to impregnate the fixed carcass with acid alcohol for a day or so before immersing it for at least a week in a 1% solution of Victoria Blue in acid alcohol. This will overstain the specimen, necessitating a period of destaining in fresh acid alcohol. Dependent upon the size of the specimen and the effectiveness of the solutions used, this differentiation will take anything from a few days to several weeks. Most small embryos should be sufficiently differentiated after three or four days. However it will require some patience and experiment to determine differentiation time for specific specimens, as the state of the stained skeleton cannot be assessed until much later in the process when the tissues are cleared.

Following differentiation the specimen must be dehydrated by treatment with 90% alcohol, followed by at least two changes of absolute alcohol, each immersion for a day or so. The specimen is now cleared in *methyl benzoate*. The fluid will render the specimen transparent within a few hours. Because it also extracts excess Victoria Blue from its tissues, it will be necessary to replace the methyl benzoate for final display. It must be stressed that methyl benzoate is harmful to health and should be handled with appropriate care.

b. Alizarin Red

Specimens are fixed in 70% alcohol for several days and then defatted by immersion in acetone for a day or two. The tissue is then partially macerated by transferring the specimen to a weak potassium hydroxide solution. The strength of the hydroxide used will be governed by the size of the specimen, ranging from say 0·5% for small embryos up to about 2% for animals the size of guinea pigs. In any case the specimen is left to macerate until the appearance of the first signs of the

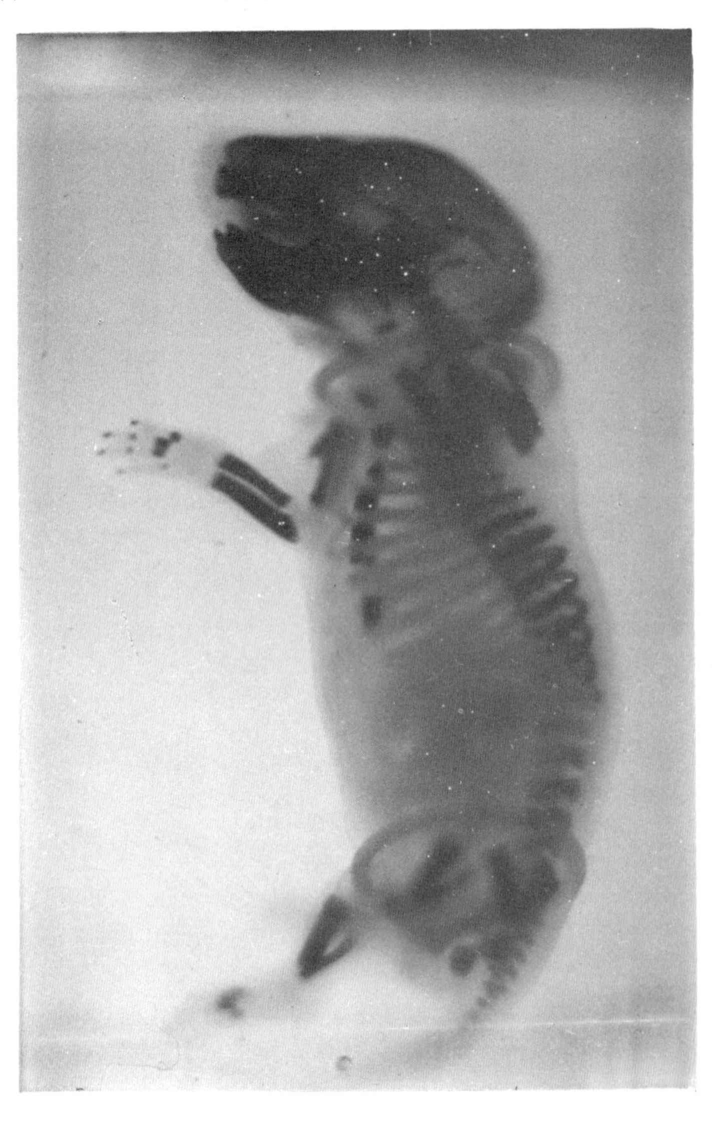

Fig. 8:5 An Alizarin Red-stained rat embryo

skeleton showing through—usually the ends of the digits. Experiment is needed to assess the correct maceration time. Over-maceration results in the specimen falling apart, while under-maceration prevents adequate clearing later. However, when maceration is sufficient, replace the hydroxide solution and add a few drops of absolute alcohol which has been saturated with Alizarin Red S. The hydroxide should assume a pale red colour. The bony skeleton (and any other tissue containing calcium) should take up the stain in a day or two. If necessary, excess stain may be removed by placing the specimen is slightly acidified (1% hydrochloric acid) industrial spirit. This acts very quickly and the specimen must be watched very closely to prevent excessive destaining.

The specimen must now be cleared. This is effected by transferring it through increasing strengths of aqueous glycerol solution, say 10%, 50%, 70%, 80%, 90% and finally two changes of 100% glycerol. Immersion in each should last for two or three days. This clearing process may be improved, however, by the following treatment *after* staining:

Place the specimen in 0·8% (w/v) solution of potassium hydroxide in 20% aqueous glycerol for a few days. Then transfer it for a day to a mixture of glycerol: 70% spirit: benzyl alcohol (2:2:1). The benzyl alcohol greatly enhances the clearing effect of the glycerol. Then return the specimen to the hydroxide/glycerol solution for several days. Now clearing may be completed with increasing strengths of aqueous glycerol, say 50%, 70% and two changes of 100% glycerol. Absolute clearing may take several weeks, or even several months, and the specimen should receive another change of pure glycerol after this time. Figure 8:5 shows a mounted Alizarin Red stained transparency.

c. *Mounting skeletal transparencies*

The preparation of a transparency renders the specimen very

delicate indeed, and the maintenance of transparency requires total and permanent immersion in the appropriate fluid. These features of the technique raise problems of mounting the finished specimen in a suitable container for display. The extreme fragility of the specimen makes it almost impossible to fix to a backing base with filament or adhesive. An effective technique is to support the specimen between the glass wall of its container and a glass plate made to fit exactly into the container. (See figure 8:6.) It is inadvisable to use a container narrow enough to support the specimen by itself, as this is likely to be top-heavy. It is important that any container used be leak-proofed, especially in the case of Victoria Blue transparencies where any escape of methyl benzoate would constitute a health hazard.

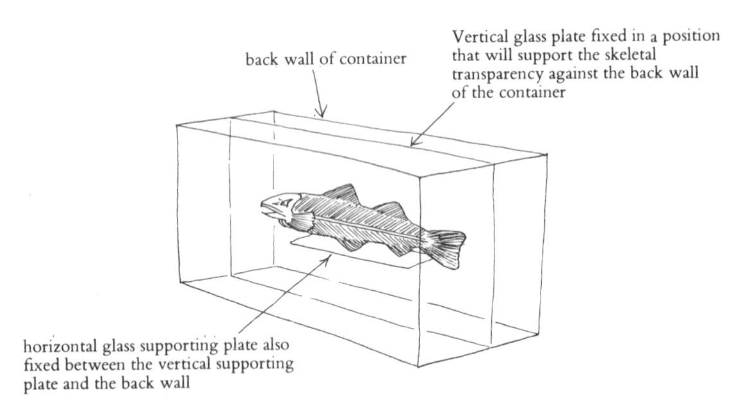

Fig. 8:6 A method of mounting the skeletal transparency of a fish

Sources of materials

All the materials, stains and other reagents mentioned in this section may be obtained from any of the Biological Supply companies listed on page 93.

References

Mahoney, R. (1966) *Laboratory Techniques in Zoology*, Butterworths.

Tompsett, D. H. (1970) *Anatomical Techniques* (Second Edition), Livingstone.

Wagstaffe, R. & Fidler, J. H. (1968) *The Preservation of Natural History Specimens* Vol. 2. Vertebrates. Witherby.

CHAPTER NINE

HISTOLOGICAL TECHNIQUES

HISTOLOGY IS THE MICROSCOPIC STUDY OF THE TISSUES OF PLANTS and animals. When fresh tissue is obtained it must be treated in such a way that it will not decay or distort, that is, it must be *fixed* (see chapter one). Many fixatives are available for histological work, the choice depending on the tissue, the stain to be used, and the features of the tissue to be observed. Some of the commoner fixatives are:

Formol Saline is a good general fixative and tissues do not usually suffer from prolonged immersion in it.

0·9% aqueous sodium chloride	90 cm³
formalin (40% formaldehyde)	10 cm³

Add a few granules of calcium carbonate to give an approximately neutral solution.

Bouin's fluid tends to prevent the hardening which occurs in some tissues fixed in formol saline. Tissues should be fixed in this for about a day.

Saturated aqueous picric acid	75 cm³
Formalin	25 cm³
Glacial acetic acid	5 cm³

Carnoy's fluid has a very rapid action, and small pieces of tissue may be treated for a couple of hours before transfer to absolute alcohol.

absolute (100%) ethyl alcohol	60 cm³
chloroform	30 cm³
glacial acetic acid	10 cm³

Following fixation, the tissue is stained with a dye that will highlight different structural and functional regions and differentiate cellular contents. Again a wide variety of stains is in use, depending upon the material being studied. Stained material must be dehydrated before mounting on a microscope slide, because the mountants commonly used do not mix with water, taking on a milky appearance which prevents microscopic observation.

Graded alcohols are usually employed, washing the stained tissue in ever-increasing concentrations of ethyl alcohol until all the water has been replaced (see chapter three). Unfortunately the alcohol will not mix with the mountant either and must itself be replaced before mounting. This is called *clearing,* and clove oil or xylene are the commonest clearing agents in use. Only now can the stained tissue be permanently mounted on a microscope slide. *Canada balsam* is widely used as a mountant and is applied to the preparation on the slide, when dissolved in xylene. This allows the balsam to flow freely over the tissue and a cover slip to be placed on top. The xylene will evaporate leaving the balsam solid and holding the cover slip firmly in place.

Techniques

Some tissues are readily available in a form that can be directly placed on a microscope slide and there treated with fixative, stains, etc. These include the following:

a. Smear preparations

This technique is suitable for blood, epithelial cells from the inside of the mouth, epithelial cells from the vaginal lining of small laboratory animals, or drops of bacterial cultures. Only the method for preparing blood smears will be described.

A small drop of fresh blood taken from the ear lobe or thumb is placed about $\frac{3}{4}$ of the way along a clean microscope slide. A clean cover slip, or another slide, ideally with ground edges, is then placed with one edge on the slide, near to the drop of blood. Holding this at an angle of about 45°, it is then carefully brought towards the drop of blood until it just touches. The blood will spread out along the edge of the cover slip towards the edges of the slide. Now carefully spread the blood as a smear over the slide by dragging the cover slip along the slide to the other end. This procedure is illustrated in figure 9:1. The smear is now quickly dried by carefully

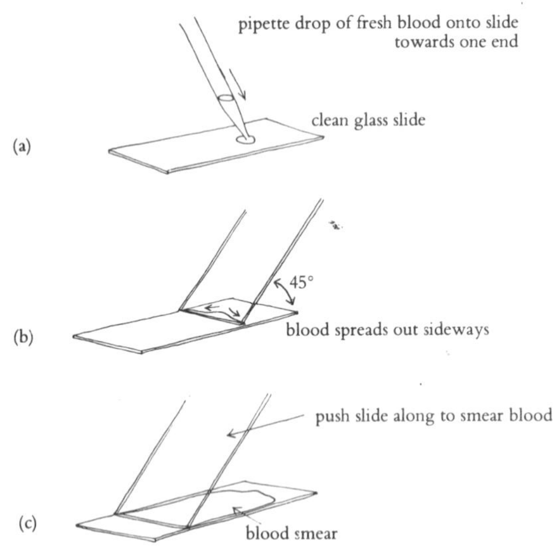

Fig. 9:1 Preparing a blood smear

waving the slide in the air. The preparation of a smear in this way must be fairly rapid, otherwise the blood will clot and be useless for normal microscopic study.

Blood smears are conventionally stained with *Leishman's stain* which may be obtained commercially, ready dissolved in methanol which acts as a fixative. Cover the smear with this stain and leave it for 30 seconds. Now add twice its volume of buffered distilled water and gently rock the slide to mix until an irridescent scum appears on the surface. Now leave the smear to stain for about 7 minutes. The distilled water is made up as follows:

M/15 disodium hydrogen phosphate	(9·47 g per litre)	61 cm^3
M/15 potassium dihydrogen phosphate	(9·08 g per litre)	39 cm^3
distilled water		2 litres

Now pour off the stain and cover with distilled water for 2 minutes to remove excess dye. Finally blot the smear dry with clean filter paper, pour on some Canada balsam, and apply a cover slip. The mounting procedure is shown in figure 9:2. Plate 8 (page 85) is a photomicrograph of stained human blood cells.

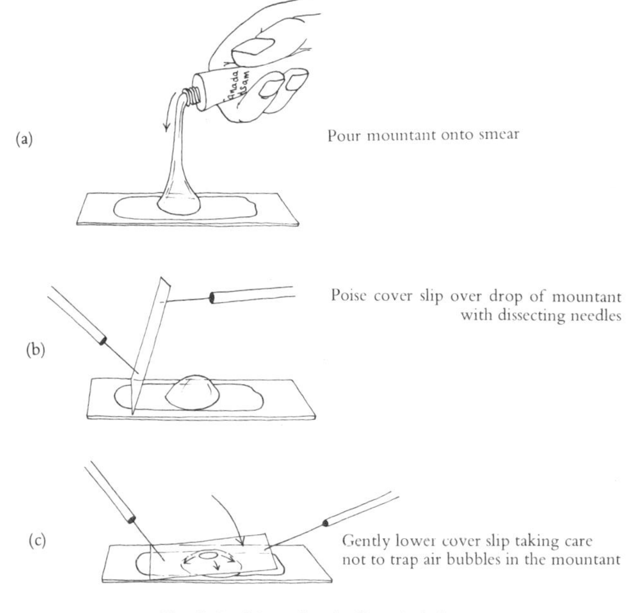

Fig. 9:2 Mounting in Canada balsam

b. Squash preparations

This technique is suitable for preparations of the seminal vesicles of earthworms, the testes of locusts, plant root tips which are described below, and the large salivary glands of *Drosophila* larvae.

Actively growing root tips of such plants as broad beans are a good source of cells undergoing *mitosis*. The growing tips of the roots should first be fixed overnight in acetic acid: alcohol (1:3) mixture. Now hydrolyse the roots by immersion

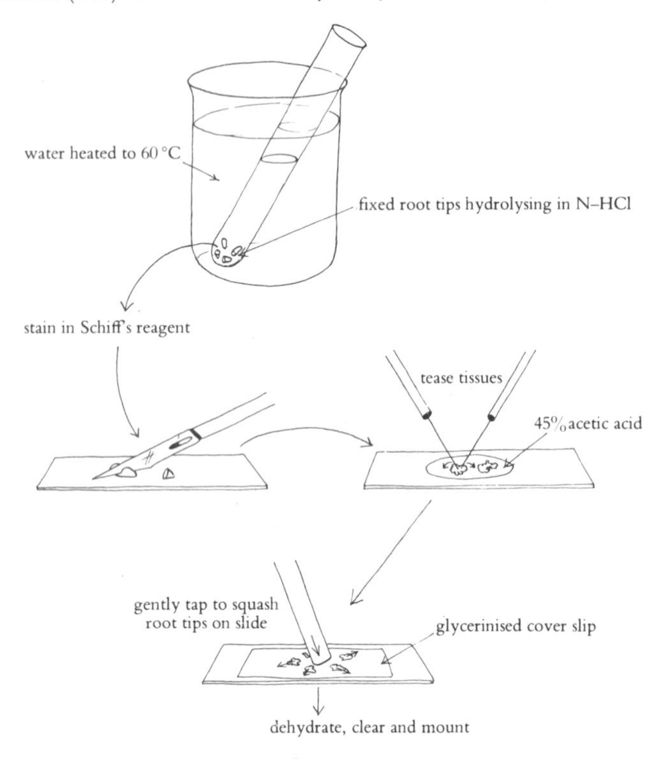

Fig. 9:3 Root tip squash preparation

for exactly *six minutes* in 1N-HCl at 60° C. Care should be taken with hot acid and this is best done in a water bath. After the acid treatment stain the roots in *Schiff's reagent* in a tightly-corked tube for about 2 hours. The root tips will take on a pink colouration. Remove the roots onto a clean slide and, with a sharp scalpel, remove the terminal 0·5 to 0·75 mm. Mount these in a drop of 45% acetic acid on another slide. Again using a scalpel and a fine dissecting needle, tease and dissect the root tips. Lightly smear a clean cover slip with glycerine/albumen to prevent the root tips sticking to it, and gently lower it onto the preparation. Now carefully tap the centre of the cover slip with the handle of the scalpel so as to squash the preparation gently, and push the cells out into a thin layer. This is completed by pressing down on a piece of filter paper placed over the cover slip, being careful not to break it.

Finally the cover slip is slowly removed and the root tip squash on the slide is dehydrated with graded alcohols (5 to 10 minutes in each) cleared in xylene and mounted in Canada balsam in the usual way. The whole procedure is shown in figure 9:3. This particular method will stain only the nuclei and chromosomes. Figure 9:4 illustrates the kind of preparation that might be obtained.

c. Teased preparations

This technique is suitable for fibrous tissues such as tendons or striped muscle. Remove a small piece of striped muscle about 5 × 2 mm from a suitable freshly-killed animal and fix it for about 10 minutes in 70% alcohol on a clean microscope slide. While fixing, the muscle is held firmly at one end with fine forceps, and the fibres of the other end are carefully teased apart with the point of a fine dissecting needle. (See figure 9:5.) Now stain the fibres in *Ehrlich's haematoxylin* for about 5 minutes. Excess stain is then removed by immersion in acid alcohol (1% HCl in 70% alcohol). The fibres must be viewed

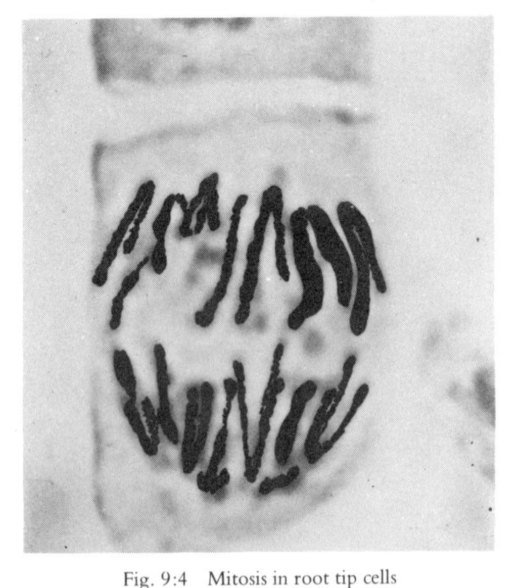

Fig. 9:4 Mitosis in root tip cells
From *Photomicrographs of Flowering Plants* by Shaw, Lazell and Foster, Longmans 1965.

microscopically during this differentiation in order to judge when it is complete. If one destains too much, the haematoxylin may be applied again, followed by further acid alcohol treatment. Stop differentiating by washing the fibres in non-acidified 70% alcohol and then immerse them in ammoniated 70% alcohol. The change in pH will turn the stain blue. In

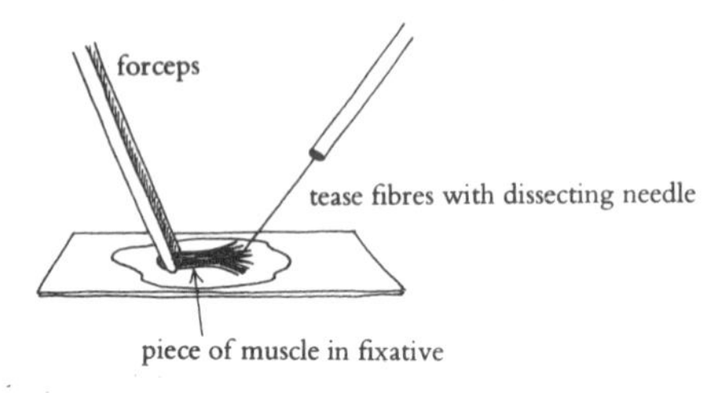

Fig. 9:5 Teasing muscle fibres

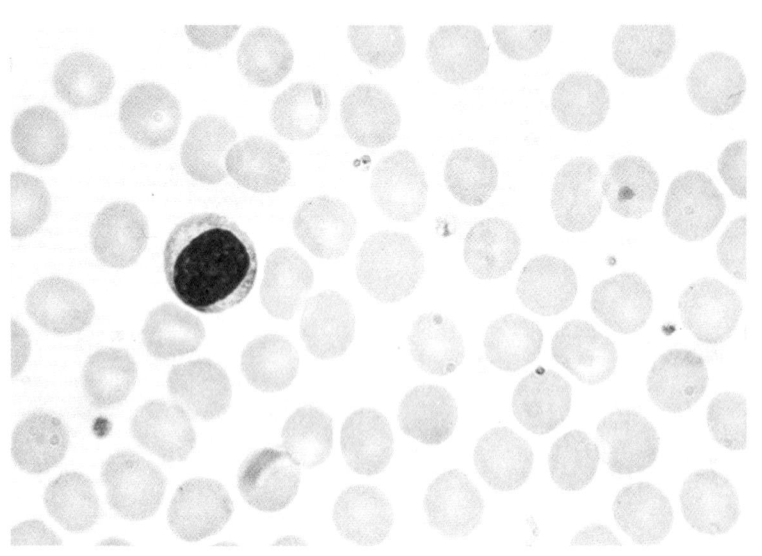

Dr. G. A. McDonald, Royal Infirmary, Glasgow

Plate 8 Human blood cells × 1200

many areas, tap water is sufficiently alkaline to cause *bluing* and may be used instead of ammoniated alcohol. Now

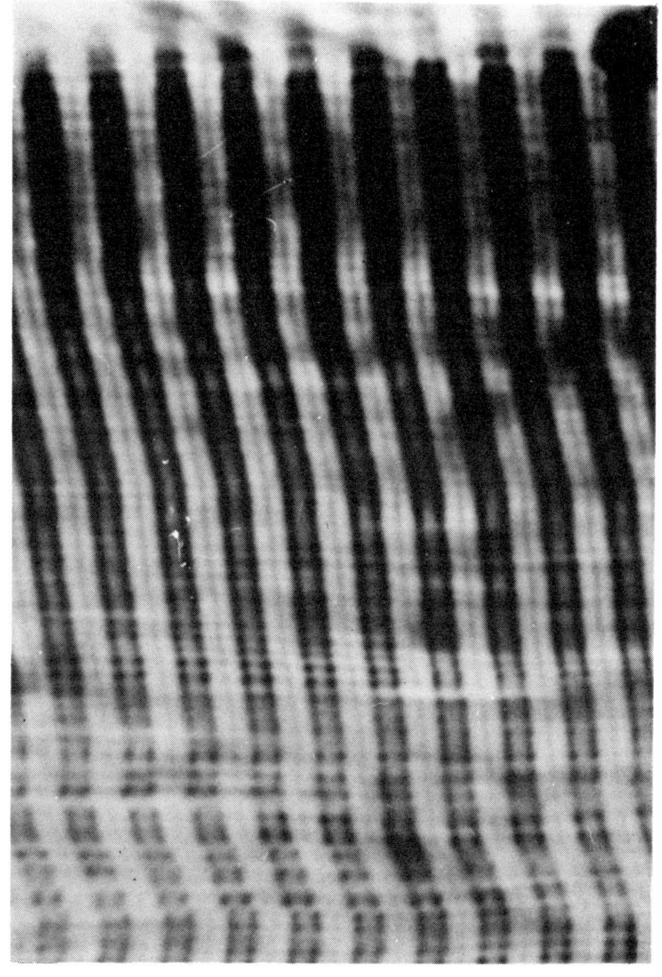

Fig. 9:6 Striped muscle fibres × 1100
From *Atlas of Histology*, by Freeman and Bracegirdle, Heinemann, 2nd edition 1967.

dehydrate the fibres with graded alcohols, clear in clove oil and mount in Canada balsam beneath a cover slip. Figure 9:6 shows stained teased striped muscle fibres.

d. Hand-cut plant sections

The internal anatomy of plant stems, roots and leaves can only be viewed microscopically if thin sections of these organs are cut prior to staining and mounting. Sectioning by hand requires practice, but good results are readily achieved. The plant material must be firmly held vertically in the left hand (if you are right-handed). It may be supported if necessary by pieces of dried pith obtained commercially or suitably shaped pieces of fresh carrot. Now hold a razor in the other hand, as shown in figure 9:7. Being careful not to cut your hand and holding the blade horizontally, draw the razor towards the end of the tissue and try to cut a thin section. It is usually best to perform this action quickly, even though a large proportion of the sections produced will be useless. If one tries to cut each section slowly, most often they will be too thick. It is rare that a complete transverse section one or two cells thick is obtained, and one should select half-sections that reveal the structures to be studied. All the sections cut should be placed in a dish of water and sorted with a camel-hair brush. Transfer good material to a dish of 70% alcohol to fix. There are several dyes that may be used to stain plant material, such as the following:

Place several fixed sections on a slide and apply a few drops of *Safranin* to stain for about 10 minutes. Now dehydrate in the usual way. Transfer the sections from absolute alcohol to *Light green* in clove oil for about a minute. This stain is made by dissolving 1 g light green in 100 cm^3 of 75% clove oil in absolute alcohol. Now clear in clove oil and mount in Canada balsam. (See figure 9:8.) At no time should plant sections be unduly exposed to the air or the staining will be impaired.

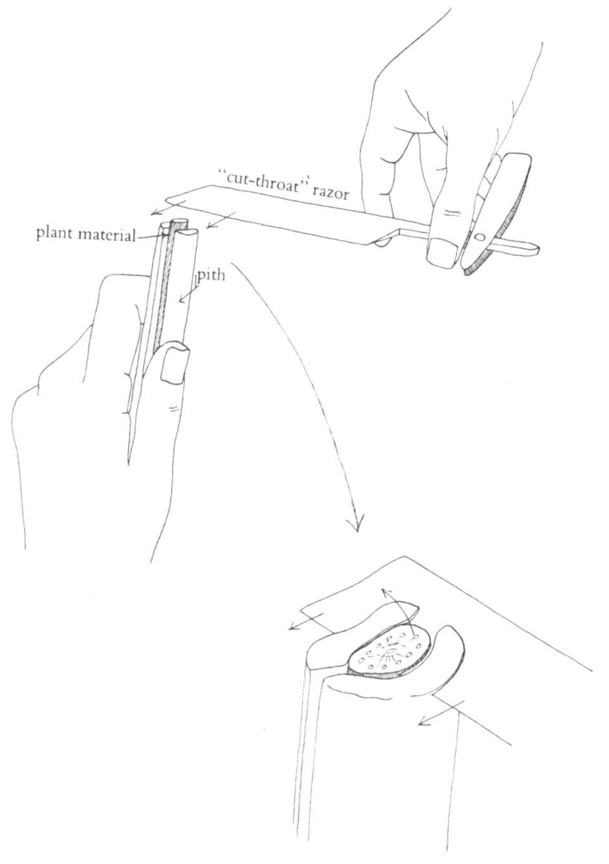

Fig. 9:7 Cutting plant sections

e. Embedded sections

Most animal tissues are so delicate and soft that adequate sections for microscopic study can only be obtained after the tissue has been impregnated with and embedded in some supporting medium. *Paraffin wax* is most commonly used for

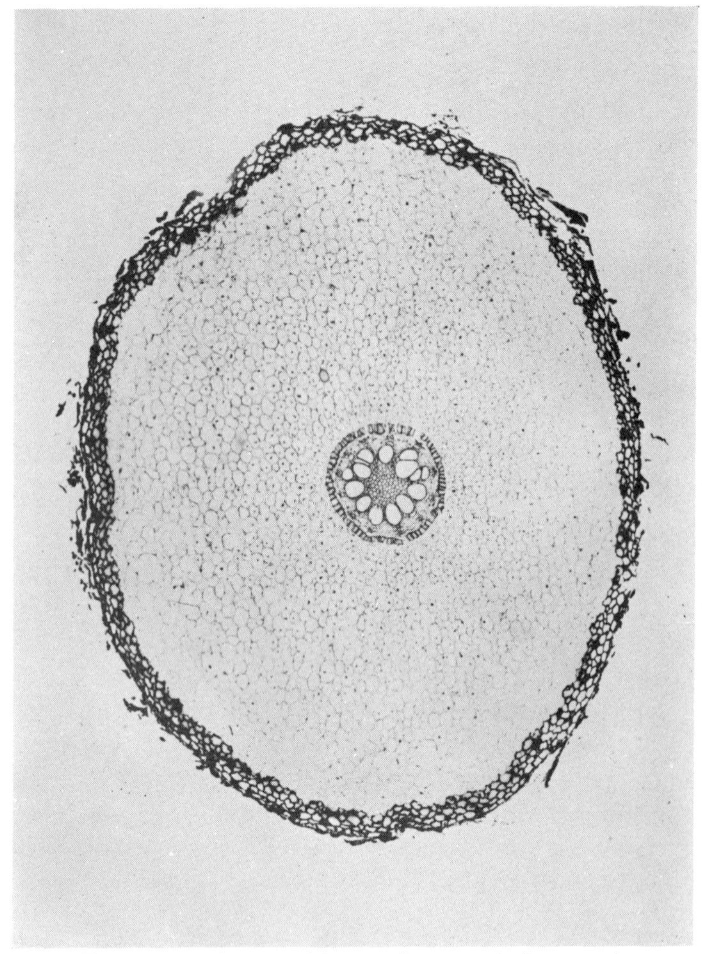

Fig. 9:8 Stained section of the root of *Iris germanica* (low power)
From *Photomicrographs of Flowering Plants*, by Shaw, Lazell and Foster, Longmans 1965.

this purpose. Fixed tissue must be dehydrated with graded alcohols and cleared before being transferred to the wax. The following technique is suitable for most organs such as liver or testes, but the pre-fixation detail is specific to the gut.

Short lengths of gut must be speedily yet carefully removed from a freshly-killed animal. The slightest delay will result in decay of the inner *mucosa*. Each piece, about 1 cm long should then be immersed in fixative, say formol saline and carefully cut open lengthwise so as to expose the muscosa to the fixative. It is best to hold the pieces of gut open by pinning them to a small cork board at the bottom of the fixing dish. (See figure 9:9.) Leave to fix for 24 hours and then dehydrate by replacing the fluid with graded alcohols, leaving in each for about 1 hour.

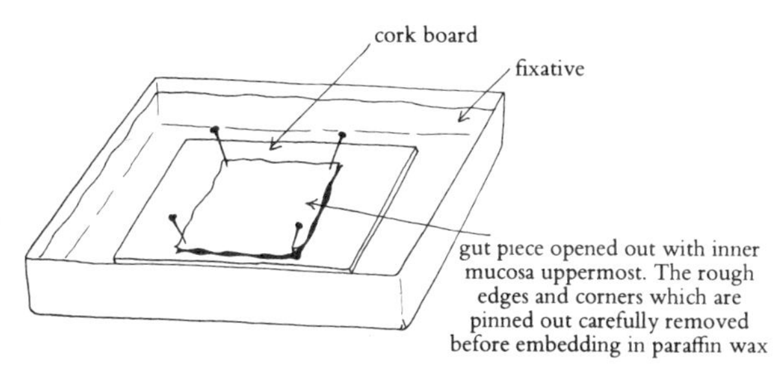

Fig. 9:9 Fixing gut portions

After two changes of absolute alcohol, clearing is achieved by an hour's immersion in xylene or clove oil. Now gently heat a suitable quantity of paraffin wax to a temperature just above its melting-point, usually 55° or 60° C, depending upon the variety purchased. Melting is easier if the wax is cut into small blocks or shavings first. A mould for the wax is made by arranging two L-shaped pieces of metal on a square of glass as shown in figure 9:10. Glycerine/albumen smeared on these will facilitate the release of the block later.

Now transfer the cleared tissue to the molten wax and leave to impregnate for 2 hours (1 hour in each of two wax baths). When this is done pour some wax into the mould and carefully

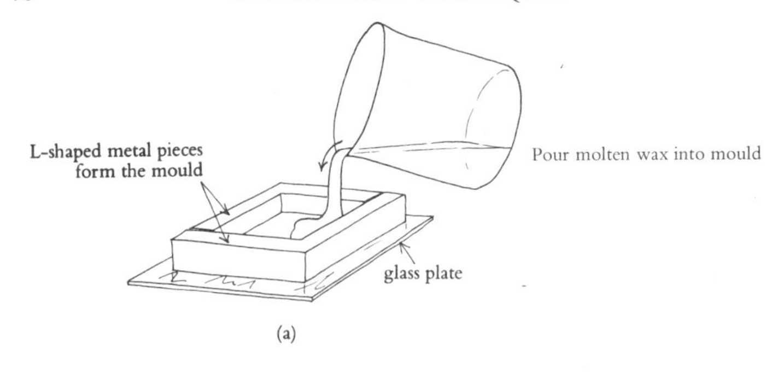

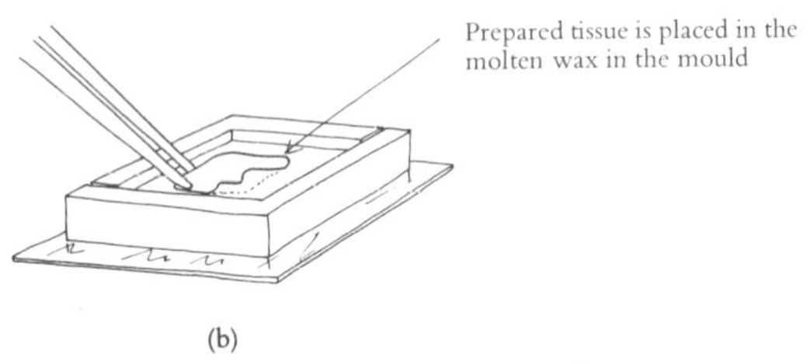

Fig. 9:10 Embedding tissue in paraffin wax

place the wax-impregnated tissue into this using warmed forceps. Align the tissue in the mould in such a way that it will be sectioned in the desired plane. (See figure 9:10.) Now gently lower the mould into a dish of cold water. This will hasten the setting of the wax block, producing smaller wax crystals. In this state the block will cut more easily than if it is allowed to cool slowly. Histology laboratories usually possess tissue-processing machines which can be programmed to transfer tissues from one solution to another automatically. These ease the sometimes laborious process of fixing, dehydrating, clearing and impregnating.

Once a tissue is embedded in wax and the block has been carefully trimmed square with a scalpel, thin sections are

Fig. 9:11 The Cambridge rotary rocking microtome

obtained with a *microtome*. (See figure 9:11.) The wax block is attached to a slightly larger wooden block by melting one surface of the wax, avoiding the tissue and pressing it firmly onto the wood. This is now clamped onto the microtome arm. When operating, this rocks up and down, advancing by the preset thickness of the sections being cut between each cycle. (See figure 9:12.) As the arm moves down, the wax block and tissue are brought into contact with a sharp blade which cuts the sections. The use of a microtome requires skill, and details of microtomy are outside the scope of this book. For further information, see the references below.

Sections obtained from the microtome are floated on water kept at a temperature just sufficient to soften the wax but not melt it. Now the sections can be straightened out with a camel-hair brush if they are wrinkled. Good sections are placed on microscope slides by the method illustrated in figure 9:13. The sections are dewaxed by immersion in xylene until only the tissue remains on the slide. A wide variety of staining

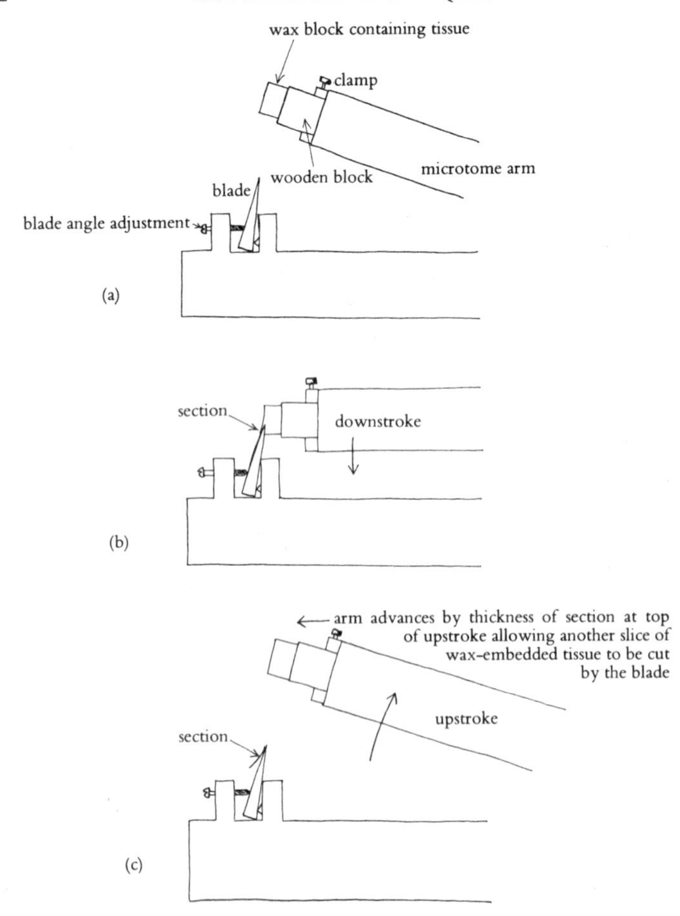

Fig. 9:12 Microtome sectioning

techniques exists for such materials, but space only permits the description of the following:

After dewaxing, hydrate through the graded alcohols (100%, 90%, etc.,) until 70% is reached and stain with *Ehrlich's*

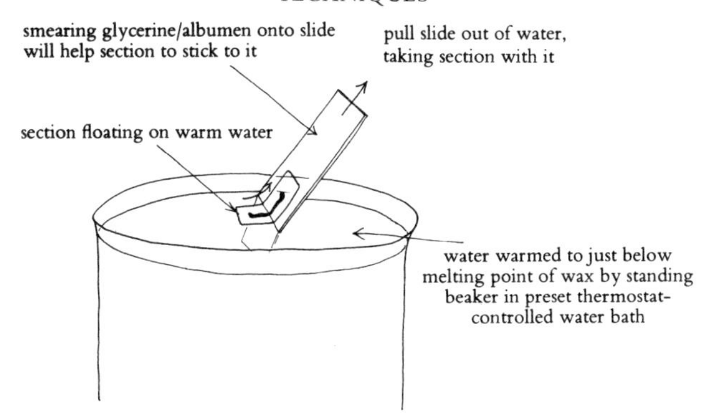

Fig. 9:13 Mounting wax-embedded sections on a slide

haematoxylin as described for striped muscle in *c* above. After *bluing,* wash the sections in water and counterstain with 1% *aqueous eosin* for 4 or 5 minutes. Now dehydrate with graded alcohols (2 minutes in each), clear with xylene and mount in Canada balsam.

Gratitude is expressed in the preparation of this chapter to Mr. C. R. Pashley, lecturer in biology, People's College of Further Education, Nottingham.

Sources of materials

All stains, clearing agents and mountants can be obtained from any of several suppliers. Among these are:

BDH Chemicals Ltd., Poole, Dorset.

George T. Gurr Ltd., Carlisle Road, The Hyde, London NW9.

Philip Harris Ltd., Ludgate Hill, Birmingham B3 1DJ.

References

BDH booklet (1972) *Biological stains and staining methods,* BDH Chemicals Ltd.

Culling, C. F. A. (1963) *Handbook of Histopathological Techniques,* 2nd. edition, Butterworths.

Fowell, R. R. (1964) *Biology Staining Schedules for First Year Students,* 8th. edition, H. K. Lewis & Co. Ltd.

Gillison, M. (1962) *A Histology of the Body Tissues*, 2nd. edition, Livingstone.

INDEX

INDEX